Quality Assurance
in
Service Organizations

Anthony DiPrimio

Quality Assurance in Service Organizations

Chilton Book Company
Radnor, Pennsylvania

Library of Congress Cataloging in Publication Data
DiPrimio, Anthony.
  Quality assurance in service organizations.
  Includes index.
  1. Service industries—Quality control.
I. Title.
HD9980.5.D52   1987   658.5'62   86-47961
ISBN 0-8019-7730-4

1 2 3 4 5 6 7 8 9 0   6 5 4 3 2 1 0 9 8 7

Contents

Preface

This book explains how to develop, implement, and administer a quality assurance and service reliability (QA&SR) program, from the initial presentation of the proposal to senior management, through every step in the development process, to the end-of-year evaluation report. I will tell the reader what to expect to accomplish with a QA&SR program, how to get the program into operation, and how to run it.

I make the following assumptions about you, the reader: That as a manager or senior officer, you are interested in reducing operating costs; eliminating customer/client complaints about poor quality or poor service; bringing in new and profitable business; keeping customers/clients loyal to your organization; and increasing efficiency and profitability. This book shows how to use a QA&SR program to accomplish these goals.

The book is organized into nine short chapters. Chapter 1 begins with an explanation of quality assurance and service reliability and describes the major components of a QA&SR program: quality measures, quality engineering, quality education, and quality management. The three dimensions of quality—design quality, production quality, and perceived quality—are also explained. Additionally, the differences among quality assurance, quality circles, and quality control are explained.

Chapter 2 explains why there is an acute need for service organizations to improve the quality and reliability of their services. The chapter also shows how a QA&SR program reduces operating costs, draws knowledgeable customers, retains customers, and supports technological improvements.

Chapter 3 describes how to get a QA&SR program accepted by senior management. A specimen copy of a proposal is included and described. The chapter also explains how to pull together a matrix organizational structure of QA teams.

Chapter 4 covers how to build positive attitudes among employees and gain their commitment to improving quality and service reliability. This chapter also explains how to use employee participation in the QA&SR program to identify conditions or problems that hinder employees from doing error-free work.

Chapter 5 discusses how to analyze the causes of errors to identify conditions and environmental situations that lead to quality and service reliability problems. This chapter also includes suggestions for using face-to-face interviews with employees to identify conditions that cause or contribute to such problems.

Chapter 6 explains how to develop a QA&SR measuring monitoring system. It offers suggestions regarding what to measure, how to gather information, how to present the information, and how to integrate the QA&SR monitoring system into your organization's management information system. This chapter also shows how to use survey questionnaires to find out what clients think about your organization's services.

Chapter 7 explains how to plan and implement strategies such as communication workshops, educational outreach programs for clients, mailings of educational materials for clients, standardized preprinted control documents and forms, training programs, and software verification checks to eliminate the underlying causes of quality problems.

Chapter 8 discusses how to document and evaluate the benefits of the QA&SR program. Sample copies of QA team minutes, reports for senior management, and a sample of a survey questionnaire that can be used to gauge improvements in client perceptions of service quality are included in the chapter.

Chapter 9 explains how to integrate the QA&SR program into the management process. The chapter also explains how to incorporate quality and service goals into the employee performance evaluation system and why doing so is important. Additionally, the chapter discusses how to use nonfinancial awards to recognize and reward QA team members and other employees who have made significant contributions to the QA&SR program. Sample copies of award forms that are useful in setting up a QA&SR award program are included.

Appendix A contains a prototype of a useful quarterly report. Also included is a comprehensive description of what a successful QA&SR program can accomplish in three months, including the benefits that may be achieved through an interdivisional communication workshop. Appendix B explains how to develop a decision support system (DSS) to monitor the progress of the QA&SR program.

I hope that this book will enable you to move confidently from the inception of a QA&SR program to its successful implementation and administration in your organization.

Acknowledgments

Lots of people helped me write this book. None of them—except for the people at Chilton and my wife—actually set out to help me write it. They just gave me the ideas and experiences that I organized into the book. I would like to thank them by acknowledging their contributions.

Phillip Crosby, Dr. W. Edwards Deming, and many other consultants/writers helped me formulate my approach to using quality assurance in service organizations. Dick Smoot, first vice president, and Ronald Watson, senior vice president, of the Philadelphia Federal Reserve Bank helped me organize, implement, and administer the quality assurance program in that bank.

Gerald Pacella, executive vice president of the Fidelity Bank (Philadelphia), was responsible for refining the QA process and introducing enhancements that greatly broadened its scope and influence. He first saw the need to incorporate productivity and training into the QA process. He also saw the need to support the QA process with a combined management information system and decision support system (MIS/DSS) to continuously monitor and report on the effectiveness of the QA program. Mr. Pacella also introduced organizational management refinements that ensured that the QA process would become an inseparable part of the management process.

Paul Krachuk, senior vice president of Fidelity Bank, was the first to introduce the QA process into his Commercial Operations Group. His support and guidance at the crucial pilot stage enabled the QA program to gain acceptance at

Fidelity Bank. Charles Angelo, vice president of the Trust Operations Group, showed how the QA process could be used in a turn-around situation. Arthur Gallagher, senior vice president of the Banking Operations Group, added refinements to the QA process to meet the demands of a large city bank's checking operation. Dick White, senior vice president of the Consumer Lending Group, added refinements to fit the specific needs of his group.

It is not possible to mention all the QA team leaders and members at the Philadelphia Federal Reserve Bank and Fidelity Bank who helped to make their QA programs work, but I will always remember them.

Eleanor, my wife, worked the hardest of all, typing through all the revisions. Her sacrifice in spending days and nights typing and in foregoing entertainment and companionship as we worked on the book will long be remembered. To her and to my children, Denise, Anthony, and Victor, I dedicate this book.

Chapter 1

What Is a QA&SR Program?

A QA&SR program is a planned, systematic way of finding and eliminating the root causes of quality problems. These causes include poorly designed processing systems, inadequate training, poor work attitudes, poor communication, poor working relationships with interconnected departments, and poor-quality incoming work from clients or suppliers.

The purpose of a QA&SR program is to provide a matrix organizational structure of quality assurance teams (QA teams) that utilizes the knowledge, experience, and drive of line personnel to identify, document, and correct quality and service problems. The QA&SR program provides a dynamic process for developing and implementing strategies to correct quality and service problems by eliminating their underlying causes.

Four major components provide the foundation for a QA&SR program:

- Quality measures
- Quality engineering
- Quality education
- Quality management of input

Quality Measures

A first step in implementing a QA&SR program is to review the quality measures in a department or division to make sure that management is provided with the kind of

1

information needed to evaluate and monitor the quality of the department's operations. Effective quality measures will give managers a sense of the quality problems in their operations and will indicate whether these problems are being corrected. In addition, such measures will enable managers to determine the impact of these problems on costs and service to clients and how these problems affect the perceptions that clients—inside and outside the organization—have of their operation.

Most service organizations do not have effective quality measures. For example, a large bank reviewed the way it measured the quality of its operations and found that no way existed for managers to learn what clients thought about the quality of the bank's services. The only measure that came close was a count of the number of complaints received. This measure was of limited usefulness, because it merely identified those services that were bad enough to motivate clients to complain. The measure did not tell what clients thought about the quality of other services, which may not have been bad enough to complain about or which may have been such a long-standing problem that clients had given up on complaining about them. The measures also did not tell management which services clients liked and which they felt should be continued or expanded. This bank also found that no counts were kept of quality related problems, such as errors made in processing work.

The thrust of the quality measures component of a QA&SR program is to formulate precise definitions of quality in terms of specific operations, such as check processing, securities processing, customer services, etc. This component also includes making sure that the client's perceptions of quality are monitored periodically.

Quality is a complex concept consisting of three dimensions:

- Design quality
- Production quality
- Perceived quality

Design quality measures the nature and variety of services offered to determine their market acceptability and their fulfillment of client needs. It also measures the responsiveness and adaptability of the organization in accommodating its clients. Examples of elements that define design quality are service packages, service schedules, types of coverages, regulations governing how services are provided, etc.

Production quality measures the reliability of the service in terms of accuracy, timeliness, and consistency. Whereas design quality measures what is offered, production quality measures how well it is offered. Of course, an error-free service that is offered on a timely basis is of little use to the client if the service offered does not meet his or her needs in terms of design quality.

Perceived quality refers to how clients view the quality of the services they receive in terms of production and design quality. An organization is in trouble if its clients perceive that the organization's design and production quality does not meet their needs. That is why all three dimensions of quality are important.

The inherent complexity of quality makes it imperative that organizations develop a comprehensive quality monitoring system that begins by specifying the standards of quality the organization wants to meet consistently.

Quality Engineering

After reviewing the quality assurance measures in an organization, managers implementing a QA&SR program must determine how, when, and where to inspect for errors. This determination process, called *quality engineering*, also consists of recording inspection data, analyzing the data, interpreting the data, and presenting the data to management.

Data regarding the performance of operative personnel should be explained to the employees involved, so that they know they are being measured and can relate to the performance measures. Employees are not likely to mind being measured if they feel the measures are fair and accurate. One of the steps in implementing a QA&SR program is to design performance measures for operative employees. Employee performance measurement data shown as trend lines charts should be displayed in the employees' work area or where they gather.

Quality Education

Quality education is an important component of the QA&SR program that is often overlooked or handled inadequately. It consists of three activities:

> Orienting people to the concepts and procedures of quality assurance,
> Training operative personnel, when necessary, on job skills and quality standards, and
> Conditioning operative employees and supervisors to be committed to meeting quality standards.

Time and effort invested in this educational process yield dividends in greater support from QA team members and operative employees.

Management of Input

To control the quality of work produced by a department, it is necessary to identify, investigate, resolve, and prevent problems with work sent into a department or division by clients. Clients can be customers who do business with the organization or other departments or divisions within the organization. The quality of work sent in for processing can be managed by:

> Establishing communication networks with clients who send in work for processing,
>
> Using communication networks to report problems and to make suggestions on how to avoid specific errors,

Establishing correction procedures and problem management systems to resolve complaints and problems reported by clients,

Establishing early warning systems that help detect potential quality problems, and

Establishing educational outreach programs that inform clients on quality related matters.

Employee Participation

QA&SR programs use the detailed knowledge of operative employees and the experience, knowledge, and authority of line managers to identify the causes of problems and to implement quality improvement strategies. The motivational power of permitting operative employees to participate in managing their operation is well known. Operative employees are encouraged to participate in the QA&SR program by identifying what they see as the underlying causes of quality problems and to suggest ways of correcting the problems.

Two benefits are gained by operative employee participation: first, because they do the work on a day-to-day basis, they are in an excellent position to identify the underlying causes of quality problems; second, and even more important, because they themselves have helped identify the underlying causes of quality problems and have participated in finding ways to correct these problems, the employees will be more likely to be committed to making the corrective changes work. Operative employee participation in the QA&SR program is an effective way to gain employee commitment to meeting quality standards and improving quality.

Equally important to the success of the QA&SR program is gaining the commitment of supervisors and first-line managers to the program. The participation of supervisors, managers, and operating officers is critical to the success of a QA&SR program. Not only does their participation ensure their commitment to the QA&SR program; it also ensures that their perspective is taken into consideration when improvements and enhancements to their department's or division's operation are planned. Furthermore, the participation of supervisors and first-line managers is necessary to get approval for changes in the way work is processed.

Quality Assurance versus Quality Circles

It may be helpful at this point to distinguish between quality assurance programs and quality circles programs, both of which rely on the participation of operative personnel. In quality circles programs, the operative employees are the primary source of ideas, suggestions, and recommendations. In a quality assurance program, on the other hand, the line officers, managers, and supervisors are the primary source of ideas, suggestions, and recommendations. The operative employees are also given the opportunity to participate by giving their ideas, suggestions, and recommendations to their supervisors.

The participative process in a quality assurance program is handled differently from that in a quality circle program in that the operative employees' inputs are percolated up through the QA team matrix structure through the supervisors, managers, and officers who make up the QA teams. Operative employee participation is encouraged and directed by these supervisors, managers, and officers.

One of the many reasons that quality circles programs have not been successful in the United States is that in such programs the role of the supervisors and managers in managing the participation of the operative employees is restricted. Additionally, ideas and recommendations from operative employees tend to be limited in scope. By contrast, the ideas and recommendations of supervisors, managers, and officers, with their broader perspective of the operation, give quality assurance programs a greater probability of achieving significant improvements in operating quality.

Quality Assurance versus Quality Control

Traditional quality control programs are product oriented—that is, they focus on setting quality standards and monitoring product output to catch defective products. Although there are clear distinctions between quality assurance and quality control, these concepts are closely related and evolved in similar ways. Quality control in this country started in 1916 when Western Electric and Bell Telephone got together to make public telephones that would be reliable and that would hold up under heavy use. The first quality control program had the following objectives:

- To plan, research, and develop the needs of a public telephone system

- To use tests to select the most reliable equipment to support the system

- To use quality control inspections during the manufacture of equipment to ensure compliance with design specifications

Using this quality control approach, Western Electric and Bell Telephone Laboratories began mass production of telephone equipment. Dr. W. A. Shewhard of Bell Telephone Laboratories introduced the use of mathematical statistics for quality control applications in the 1940s. Subsequently, the use of statistical quality control applications gained wide acceptance in the auto industry.

Quality control grew in importance and visibility in the manufacturing sector thoughout the 1960s and 1970s. Then, in the 1980s, public interest in quality exploded. Senior management became acutely aware of the public's assertive demand for high quality and service reliability. To meet this public demand for better quality and reliability, both product-producing companies and service-providing companies began to develop programs to educate their employees to the importance of producing reliable products and reliable services.

Quality control programs establish product specifications in terms of characteristics such as reliability, serviceability, and maintainability. The programs are engineering-oriented, with their foundations in statistical sampling techniques, measuring devices, inspection sampling, process-control tables, and frequency distributions. Engineering specifications for a product are set in terms of dimension and weight tolerances, and specifications for reliability under various conditions are then added.

Quality control programs in the production of tangible products are as old as mass production, and they serve an essential function. Such quality control programs, however, do not work for service organizations.

When compared with quality control programs, quality assurance programs are seen as process-oriented, focusing on eliminating the causes of recurring quality problems. As mentioned previously, these programs use participative management techniques that involve the line officers, managers, and supervisors as well as operative employees. Instead of focusing on a product, quality assurance programs focus on the process that provides the service—for example, check processing, municipal bond and coupon processing, or policy endorsement processing. These programs look for conditions such as faulty system design features, training inadequacies, attitudinal problems, etc.—anything that is causing a quality problem. Analytical methods are used to find the causes of quality problems; in addition, operative employees, supervisors, managers, and departmental officers are encouraged to identify what they feel are the causes of such problems.

Uniqueness of Service Organizations

One of the major barriers to developing quality and service reliability programs for service organizations has been management's failure to realize that the needs of service organizations are different from product-producing companies. Because of this basic difference, the methods that work for product-producing companies do not work for service companies. Like social science researchers who tried unsuccessfully to adopt the research methods of researchers in physical science, service-company managers tried to adopt the quality control methods of product-producing companies, with no success.

There was no scarcity of quality control specialists burning with eagerness to tap the wealth of service companies that were awakening to the need for programs to improve the quality and reliability of their services. However, service organizations need methods designed especially for them, not some product-oriented engineer's modification of a mathematical-statistical quality control application.

The irreconcilable differences between a product-producing company and a service organization are summarized below:

1. In service organizations, there is no product with exact specifications to examine, measure, weigh, and test for functionality. Services are intangible—that is the critical distinguishing factor that renders product-oriented quality control methods impotent in a service environment.

2. No inventory control can be established over services. Services cannot be stored, and requests for service cannot be back-logged if client satisfaction is to be maintained. In this sense, services are extremely perishable.

3. Most service organizations have to successfully contend with a strong client presence in providing the service. This adds to the demands and constraints on the service-providing process.

4. The process of providing services requires a complex, highly efficient delivery system that is time-sensitive and user-friendly.

5. Because of the dominant nature of the client's presence in the service delivery process, it is difficult to

establish concrete, objective measures to evaluate the service delivery process. Because the client's views about the services are subjective, a performance evaluation system must be able to capture a broad range of client satisfaction information and convert this to useful statistics from which logical inferences can be drawn.

For these and other reasons, service industries need quality assurance and service reliability programs specifically designed for their situation. The QA&SR programs for service organizations must cover every dimension of the service transaction, from the preparation of the service to the behavior of the person delivering the service.

Suggestions for developing our custom QA&SR program

Suggestions for developing our custom QA&SR program

Chapter 2

Why Service Organizations Need QA&SR Programs

There is an acute need for service organizations to improve the quality and reliability of their services, because the people who pay for these services today are more knowledgeable and more assertively intolerant of unacceptable quality than in previous years.

This is true of all service organizations, private sector and government sector, profit and nonprofit. The demand for better-quality products that started with Japanese imports has spread to service organizations: banks, insurance companies, government agencies, and health care facilities. Consumers, taxpayers, directors, administrators, and auditors—everyone has become more conscious and demanding of high quality.

This interest in quality improvement is showing up in many ways:

> Membership in formal quality related organizations has more than doubled since 1977.
>
> More firms, product and nonproduct, are establishing full-time quality control and quality assurance positions, and many companies are implementing extensive quality programs on a company-wide basis.

Another factor that makes quality improvement critical is the high cost of employee salaries, benefits, and energy. With such expenses, an organization cannot maintain its competitive position or stay within its budget if employee time is wasted redoing work that could have been done right the first time. Jobs that consist of spending large amounts of time cataloging, correcting, and redoing work must be eliminated.

Service organizations that have started QA&SR programs are finding that their programs enable them to accomplish the following goals:

- Eliminating the unnecessary costs of catching, correcting, and redoing work that should have been done correctly the first time

- Improving the perceptions clients have of the quality of the services they are getting

- Improving the quality and lowering the cost of service by improving the quality of work coming into each department for processing

- Drawing and retaining knowledgeable clients who demand high-quality, low-cost services

- Developing services that are custom-fitted to customer needs

- Establishing an operating environment conducive to high-tech processing methods and technology

Organizations that still have not moved to develop and implement their own QA&SR programs are finding that they are losing clients to competitors with successful programs. Some of the major reasons why service organizations need QA&SR programs are presented below.

Reducing Operating Costs

"Chicago Bank Saves $262,000." Does a quarter of a million dollars in documented cost savings sound like something that would help your organization? The First National Bank of Chicago realised savings of that amount in operating costs. Lawrence C. Russell, senior vice-president of the First National Bank of Chicago, stated at the American Bankers Association National Operations/Automation Conference in Miami that he estimates that 10 to 20 percent of a bank's operating costs result from errors made in service products departments. He added that his bank saved $262,000 in less than one year through quality improvement programs.

Want more proof that quality assurance programs lower operating costs? Let me introduce another expert witness: Phillip Crosby, former vice president of quality assurance at ITT and author of *Quality Is Free* (McGraw-Hill, 1979). He identified the following operating costs as ones that can be reduced or eliminated:

- Rework costs such as personnel, overhead, and administrative costs of tracing errors, making adjustment entries, and other tasks involved in correcting work that was done incorrectly

- Complaint-servicing costs, such as those incurred when employees spend time answering customers' inquiries about errors or service failures

- Inspection costs of monitoring outgoing work

- Software correction costs, such as the expense of making programming changes

These are just a few of the many avoidable costs that can be held to a minimum by a QA&SR program.

But aren't there costs for developing and administering a QA&SR program?, you may ask. No. Preventing errors is part of everyone's job: operating employees, supervisors, managers, CEO's. The personnel costs and associated overhead of people working on quality assurance activities is nil, because these are the same people who are responsible for producing, designing, or managing work output. Quality assurance and service reliability are an integral part of managing the work process. That is why Phillip Crosby says quality is free. It is part of what operative personnel and managers are already paid to do.

Chapter 6 explains how to establish the cost of quality problems and service failures and how to monitor and report these costs. Chapter 7 explains techniques for correcting the causes of costly quality and service problems.

Improving Client Perceptions

The perception that clients have of an organization's service quality determines whether the organization will grow and prosper or whether it will lose its client base and fail. A QA&SR program enables an organization to determine what its clients think about the quality of the services they are getting. This is done by questionnaire surveys, meetings with representative groups of clients (see the discussion of focus group meetings in chapter 6), and reviews of customer complaint logs.

QA&SR programs help to shape the perception that clients have of an organization's quality in a number of ways:

Clients are informed that the organization has a QA&SR program and is committed to maintaining a high level of service quality.

Clients also receive visits designed to help them improve the quality of the work they send in for processing (see chapter 7 for guidelines on how to use a quality outreach program).

In addition, seminars are held on quality improvement topics; such seminars both improve the quality of incoming work and foster the image of an organization committed to quality.

An organization can help shape its image as a quality-committed organization by giving the QA&SR program high visibility in reports released to the public, such as the annual report, directors' report, marketing reports, and company prospectus.

The image of a quality-committed organization can also be fostered by articles about the QA&SR program published in trade journals. Someone associated with the QA&SR program—for example, the head of the program or a senior operating official—can write an article describing the program. This article can be submitted for publication by a trade journal such as *Bank Systems & Equipment, Bank Administration, Best's Review,* or *Bottomline.* This is an excellent way to get favorable publicity for your organization.

Yet another way to use a QA&SR program to get favorable publicity for your organization is to have someone associated with the program or a senior officer of the organization appear as a speaker describing the QA&SR program at industry conferences and meetings. Often these conferences and meetings are covered by trade journals, which again helps build your organization's image as a company committed to quality and service reliability.

Improving Quality of Input

QA&SR programs enable an organization to lower costs and improve the quality and reliability of service by improving the quality of work submitted by clients or other divisions for processing. In fact, improving your organization's quality and efficiency starts with improving the quality of the work sent in for processing. The importance of working with suppliers to improve the quality of the inputs in the manufacturing process is well known. Certainly it is clear that the end product in manufacturing is seriously affected by the quality of the input components.

This is just as true in the service industry. For example, if mistake-filled checks, securities, or insurance claims are sent in to your organization for processing, this increases the probability that more mistakes will be made during the processing cycle. Ultimately, these mistakes will affect the quality or accuracy of the finished work sent to the client.

The quality of incoming work to be processed affects both the processing cost and the throughput rate. For example, a manifest document accompanying a shipment of bond coupons sent by a broker to a financial institution for processing listed 100 coupons totaling several hundred thousand dollars. The clerks opening the envelopes found only 99 coupons. A frenzied search was made of the area, auditors were alerted, and the contents of waste containers were examined paper by paper. After spending hours of expensive time, the supervisor called the broker, who searched his work area and found the missing coupon. The coupon was missing because of clerical error. Total cost to the financial institution was $11,000 in wasted, unrecoverable time, not counting the cost of the two auditors who helped in the search. And this may happen as often as twice a month.

Another example involves checks sent in for processing that tellers or other employees have encoded with an incorrect amount, incorrect paying agent, or other error. These errors cause the amount totals for clients to be out of balance. Settlement clerks must trace through hundreds of checks and tapes to find the errors in order to settle.

Such problems occur often enough to keep an entire settlement department of 23 employees and a supervisor working on three shifts. The cost of catching and correcting errors is over $650,000 a year, not including rent and overhead. In addition, an equally large adjustment department finds, corrects, and makes adjustments for errors not corrected by the settlement department. That makes a total cost of 1.5 million dollars a year to catch, correct, and adjust errors.

How many of your employees spend hours of time catching your client's errors and the errors made by your processing people? The elimination of errors. more than half of which are in incoming work from clients, has a real payoff and a substantial impact on the bottom line.

Very few service organizations set standards for the work sent in for processing, because they feel the client would not care to have their work subject to standards. This is not true, however. A smart client would quickly see the advantage of working with a service institution whose price schedule reflected the quality level of a client's work. For example, a smart client would consider and agree to meet quality standards for work sent in for processing if offered a price incentive.

In some financial institutions, one price is charged for checks that can be run through high-speed reader/sorter equipment and a higher price is charged for checks that have to be sorted by hand or using lower-speed equipment. It is simply an extension of that idea to have a price schedule that reflects the quality of work. The higher the quality—the fewer the errors—the lower the price.

Improving the quality of incoming work requires monitoring it and giving the client frequent feedback. The feedback can be just a phone call to the manager of the operation that sends the work in or an invitation for that manager to visit the processor. During the visit, the manager can meet the people who process the work he sends in. This is a perfect time to talk about quality problems in the work that is submitted.

Attracting and Retaining Knowledgeable Clients

QA&SR programs enable service organizations such as banks, insurance companies, and brokerage houses—which operate in competitive environments—to differentiate their services from the almost identical services offered by similar organizations. Soap and toothpaste companies have been doing this since advertising executives first learned the trick, even before radio commercials. The strategy is to differentiate your organization's services based on quality so that clients perceive it as unique. It is an effective way for a service organization to attract clients who demand high-quality services—especially when the price of the high-quality service is the same, or lower, than that for the poor-quality services offered by a competing organization.

Does this sound contradictory: higher-quality services being offered at prices equal to or less than poorer-quality services? Why should it? The first Japanese cars were better engineered and better built than comparable American cars, and they were priced below American cars. It was only when Japanese marketing executives realized that Americans would gladly pay more for better quality that the Japanese began to load on the extras, and then only at the urging of American car dealers.

However, there is even more convincing evidence that high quality and low price can go together. Studies by Dr. Robert Buzzell and Dr. Lynn Phillips showed that product quality exerts a significant positive influence on market share, which in turn lowers direct costs as a result of scale economies ("Successful Share-Building Strategies" by Robert Buzzell: *Harvard Business Review* 1:18, 135–140; and "Product Quality, Cost Position and Business Performance" by Lynn Phillips: *Journal of Marketing*, Spring 1983, 26–44.) These studies concluded that the attainment of high quality and the pursuit of low direct costs are not incompatible business strategies.

As important as price comparisons may be, it is perhaps even more important that presenting an image of superior quality builds client loyalty. Differentiation based on the perception of superior quality makes clients less price-sensitive, thereby protecting your organization from losing clients to competitors for small price differences. And if the perception of superior quality helps retain clients, it also helps attract new clients.

The most compelling reason why service organizations need QA&SR programs, however, is to keep from losing clients because of poor quality and poor service reliability. Client dissatisfaction over poor service quality is the number one reason for clients changing service organizations. This fact has been documented and reported by the Greenwich Associates, an independent research firm, in a study of leading banks and other financial institutions. In their 1985 study, the firm surveyed a statistically significant population of operating officers from financial institutions of various sizes. Their most important finding was that operating officers select a service organization on the basis of the following attributes: price, overall quality of operation, and specific quality attributes, such as accuracy of service, speed of service, and responsiveness to requests. Every one of these attributes, including price, is improved by a QA&SR program.

The same study also reports that when price differentials are small, the major reason for switching from one service provider to another competing provider is perceived differences in quality or service reliability. And in many cases a somewhat higher-priced competitor is selected over a lower-priced competitor, if the quality and service reliability of the higher-priced organization is perceived as being significantly better.

The findings of the Greenwich study were corroborated in focus group interviews conducted by the Federal Reserve Bank of Philadelphia with senior executives from Third District banks. These focus group interviews centered on the question of the importance of service reliability. The senior executives who attended were both providers (vendors) and buyers of banking services. They indicated that service reliability was extremely important in selecting a provider of services or marketing their services to clients. They defined service reliability in terms of: consistency, availability, performance, accuracy, response time, and expectations met.

Identifying User Needs

It is not enough to provide a service with a high degree of reliability. Even when a service is provided flawlessly, it may not be enough to ensure that clients are retained. Why not? Because clients can be wooed away by a competitor who offers a service or group of services that better suits the client's needs.

A QA&SR program investigates the design quality and suitability of your organization's service line. This is done by questionnaire surveys, focus group interviews with clients, feedback from marketing representatives, and feedback from line personnel who interact with clients. Once you have ascertained your clients' perceptions of the suitability of your current services for their needs, your QA&SR program, through the QA team structure and its problem-solving process, can work to modify your service line to provide services that exactly suit the needs of your clients.

Although it may at first appear that determining the suitability of the organization's service line to meet client needs is a marketing function, this is not necessarily the case. The suitability of the service line is very much a part of the quality of a service. A service that does not exactly meet the client's specifications is just as much a quality problem as a product that does not meet a manufacturing client's product specifications.

For example, if a bank needs an accounting document by 10:00 in order to provide critical account balance information to its corporate clients before noon, and the financial institution the bank depends on to provide this accounting document by 10:00 chronically fails to get the document to the bank on time, then this service does not meet the bank's needs. Understanding that getting the document to the bank by 10:00 is the essential feature of the service—and meeting this delivery deadline—are quality assurance considerations.

Once the QA team knows that the client needs the accounting document by 10:00, they analyse the situation to find out why the document is not being delivered on time. They may find that the information needed to produce the document is not available from one department until 4:00 a.m. and that the program that produces the document is not scheduled to be run by the computer services department until 8:00 a.m. By tracing the document-production process from the beginning input point to the delivery, QA teams find ways to reschedule the processing steps to get the document to the client by 10:00. Often this requires a good deal of rescheduling and changes in priorities. The QA teams are ideally suited to work this out. If it is important to the client, the QA team works to provide the service.

Another example is a client who would find it more efficient to transmit billing information directly from its computer to the computer of the financial institution it relies on to process this information. Unfortunately, the financial institution cannot accommodate direct computer-to-computer transmission. In this example, the client must rely on less efficient methods, such as paper documents or computer tapes, to transmit the information.

Identifying that the client and the financial institution would benefit from computer-to-computer transmission and making this recommendation to management would be part of a QA&SR program. In this example, the QA teams might find that computer-to-computer origination of data cannot be offered to clients because of computer-line capacity problems. If so, the QA teams in the computer services department would be ideally suited to determine what is needed to provide the capacity. Perhaps priorities could be changed or more computer lines could be added in order to provide the service.

The QA teams, working with the account managers and marketing officer, are best qualified to make sure that the services offered to clients are the ones that best fit their needs.

Supporting High Technology

Almost all service organizations have back-room processing operations that evolved and grew larger over years. These processing operations are rarely studied from an engineering standpoint to determine if they are working as efficiently as possible.

One of the objectives of a QA&SR program is to analyze processing procedures and layouts to make them efficient. QA&SR programs often lead to the redesign of processing systems. Part of the redesign may require upgrading equipment and introducing new computer systems or enhancements to existing systems. The QA&SR program identifies problems with the present systems and devises strategies to make sure these problems are prevented by the newly redesigned systems.

Two examples, one from a securities processing department and one from a check processing department, should give you some idea of how a QA&SR program helps provide a smooth transition to a high-technology processing system.

In the first example, a financial institution wanted to upgrade their municipal securities processing department from a labor-intense manual system to a computerized operation. The department was staffed by long-term unit heads and clerical personnel, most of whom had more than twenty years of service. The manager was a very competent woman, newly appointed to the position. The vice president of the department had wanted to upgrade the processing system for some time but was unsure how to proceed. He was concerned that the unit heads and clerical personnel might not be able to make the transition effectively.

The vice president met with the institution's director of quality assurance and service reliability, and with his help a QA team was formed in the department. The QA team consisted of the vice president, the manager, two supervisors, and the director of quality assurance and service reliability, who served as an advisor on all the bank's QA teams.

After the QA team members were briefed on how the QA program would function (explained in chapter 3), they focused their attention on the problem of upgrading their labor-intense manual system for processing municipal bonds and coupons.

In simple terms, the operation consisted of recording in a log all envelopes containing municipal bonds and coupons sent in by commercial banks. Then routing clerks, relying on memory or by checking a complex routing directory, identified the bank acting as the paying agent for the municipality. The clerk then made up new envelopes and recorded an entry in the log showing the dollar amount, a description of the bond or coupon, and the name of the paying agent. Another clerk entered a credit entry for the sending bank and a debit entry against the paying-agent bank. These debit and credit entries were made in the financial institution's accounting system and later appeared on the sending bank's accounting statement and the paying-agent bank's accounting statement. All of these tasks, with the exception of the preparation of the accounting statements, were done manually.

The QA team studied the entire processing operation and identified the following recurring errors and problems:

- Selecting the wrong paying agent for a bond
- Putting the bond or coupon in the wrong envelope
- Failing to log in or log out a bond or coupon
- Entering the wrong debit or credit amount
- Charging or debiting the wrong bank (many banks have somewhat similar names)

The QA team considered a wide range of ways of redesigning the processing operation before deciding to automate the process using a personal computer (PC) and a computerized system developed by a large software development house.

The new manager and the long-term supervisors, as members of the QA team, helped research ways to upgrade the old, labor-intense processing operation. They were the ones who recommended the use of a PC and software system to automate the manual processing operation. Because the manager and the two long-term supervisors helped bring about the change and learned how the computer system would work, they did not feel threatened. They did not resist the change from the manual system, which they had developed and worked with for more than twenty years, to a new system that required them and their clerical personnel to sit at a PC and enter data. They were shown how to use the PC and its system, and they in turn trained their clerical personnel, who also accepted the change.

It is impossible to say whether the same smooth transition could have been made without a QA team structure. What is clear is that the transition was made with complete acceptance by older, long-term personnel. These were the same people who had not shown any enthusiasm for automation when a feasibility study had been made two years earlier, when the same vice president had decided to hold off on automation.

The second example is from the check processing department of the same organization. This department's problem involved processing nonmachineable checks—those that could not be run through high-speed processing equipment that reads magnetically encoded information, such as the bank number and the amount of the check, and records it directly into the bank's computerized accounting system. These nonmachineable checks had to be processed through old, inefficient equipment.

The senior officer in charge of the check processing department met with the director of quality assurance and service reliability and explained that new, state-of-the-art equipment was being tested. The new equipment was designed to process nonmachineable checks in the same way as the high-speed processing equipment.

The senior officer wanted to know if the QA&SR program could be used to:

> Help plan the software system to make it more efficient than the system it was replacing, and
>
> To condition the supervisors and operative personnel to accept the new equipment and software.

A QA team consisting of the senior officer, the manager, the supervisor, and the director of quality assurance and service reliability was formed. This team followed the same approach as was used in the municipal securities processing department:

- Identifying all recurring errors

- Identifying error causes

- Redesigning the processing system and procedures to eliminate these error causes

In addition, the QA team worked with the computer systems design people to develop a software system that eliminated the causes of errors and other quality problems that existed in the old system. The team also requested and helped design enhancements that provided better quality control over the data to be entered into the new system.

When the new equipment and system was installed, after

the customary testing and debugging was completed, management was pleased to find that operative employees were quickly trained. Implementation moved more smoothly than expected, with little or no drop in productivity or increase in errors. This smooth transition was attributed to the inclusion of the supervisors and operative employees in planning for the implementation of the new system.

These two examples suggest ways that a QA&SR program can help a department upgrade its processing operation, making it more efficient while improving its quality.

Suggestions for developing our custom QA&SR program

Chapter 3

Starting a QA&SR Program

Presenting Quality Assurance to Management

In a bank or company that does not have a QA&SR program, someone—a manager or department head—must take the initiative and present the QA&SR concept to executive management. This chapter assumes that you have heard about QA&SR and want to persuade senior management to develop and implement a QA&SR program in your organization.

The first and most critical step in getting a QA&SR program started is to persuade executive management that the program will solve the organization's quality and service problems and will reduce costs resulting from poor quality. It may be necessary to make executive management aware that poor quality is the underlying cause of many management problems, such as high unit costs, loss of business to competitors, marginal results in attracting new business, high volume of client complaints, and difficulty in meeting commitments.

The best way to persuade executive management that a QA&SR program can correct these problems is through an oral presentation and written proposal.

At the initial meeting with executive management it is helpful to use a slide presentation to explain the QA&SR program. The slides shown in this section will help you to prepare your presentation.

Proposal for Executive Management

In addition to the presentation, it is necessary to provide a written proposal for executive management. The proposal should identify the organization's quality problems, for example, high unit costs, high error ratios, client dissatisfaction, loss of business, difficulty attracting new clients, delays in processing work, slow throughput rate, bottlenecks, and high overtime caused by quality related problems.

Next, the proposal should explain the QA&SR program: how it works; how to implement it; how to administer it once it is implemented.

On the following pages, a prototype of a proposal is presented to help you prepare yours. Your proposal should clearly state the objectives or goals of the QA&SR program: to improve quality and service reliability by eliminating the underlying causes of these problems and substantially eliminating the costs associated with correcting and redoing work.

Exhibit 3.1: Slide 1

Quality Assurance & Service Reliability Program

- What is it?
- Why do we need it?
- What can it do?
- How does it work?
- How do we implement it?

Slide 1 Dialogue

Summarize briefly the points you intend to cover in this presentation (see the slide for more information).

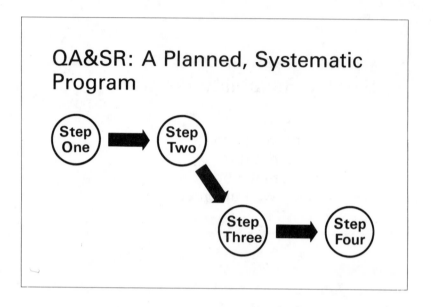

Slide 2 Dialogue

A quality assurance and service reliability program is a planned, systematic approach to quality improvement that uses group processes, analytical techniques, and operations research methods to find what causes people to make mistakes or to fail to provide good service.

Slide 3 Dialogue

The QA&SR program looks beyond errors and service failures to find the underlying causes. It searches for defects in work-processing systems, inadequate checks or verification procedures, or anything else that contributes to errors or service failures.

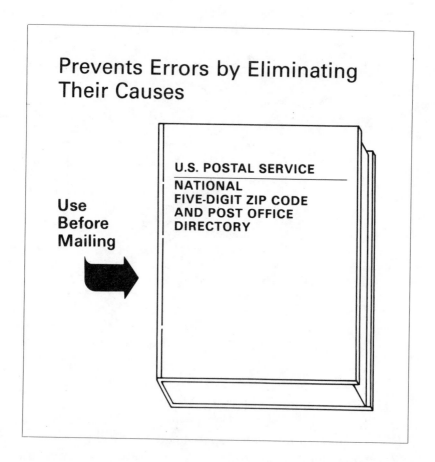

Slide 4 Dialogue

After the program identifies the underlying causes of quality problems, corrective strategies are implemented to eliminate the causes and prevent future problems.

Exhibit 3.5: Slide 5

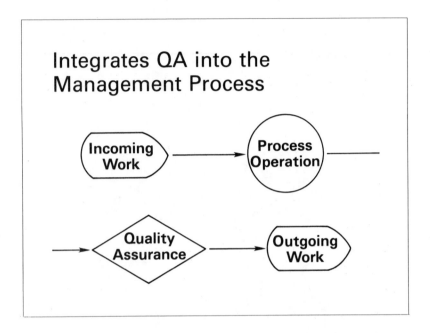

Integrates QA into the
Management Process

Incoming Work → Process Operation

Quality Assurance → Outgoing Work

Slide 5 Dialogue

This program integrates itself into the line management process. This is because the QA teams used to identify and solve problems are made up of line management people. The program is proactive in that it not only catches errors but prevents them by eliminating their underlying causes. Where necessary, the program will train employees in methods of error reduction.

Reduces Errors
Reduces Correction Time
Reduces Aggravation and Stress

Slide 6 Dialogue

Companies that have implemented similar QA&SR programs have found they reduce the number of errors made by employees, the time employees spend catching and correcting errors, and the aggravation and stress caused by errors or service failures.

Exhibit 3.7: Slide 7

Slide 7 Dialogue

Companies have also found that relationships with clients have improved and new clients have been brought in since their QA&SR program has helped provide error-free work and fewer service failures.

Exhibit 3.8: Slide 8

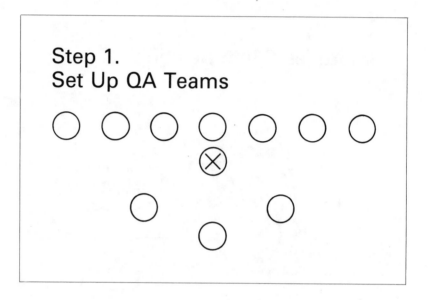

Slide 8 Dialogue

The QA&SR program is implemented through a series of well-planned steps. The first step is to set up quality assurance teams (QA teams). Each team is made up of the officers, managers, and supervisors of the department on which that team will work. The function of the QA teams is to administer the program and implement the corrective strategies developed by each team.

Exhibit 3.9: Slide 9

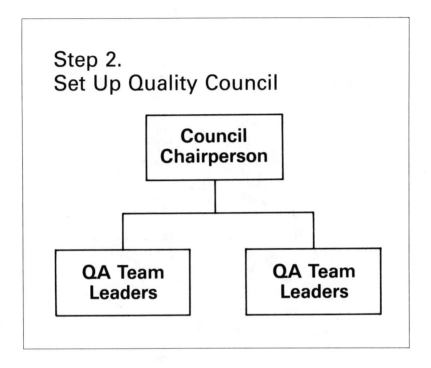

Slide 9 Dialogue

The next step is to form a quality council made up of all the QA team leaders and chaired by the senior departmental officer. The council provides a forum for briefing senior management and getting direction and feedback.

Slide 10 Dialogue

The third step is to encourage operative employees to participate in the program. This is done by explaining the program to them and asking for their support.

Step 4.
Diagnose Quality and
Service Problems

- Interviews
  - Analyses of Incoming Work
    - Analyses of Client Complaints
      - Analyses of Environmental Factors

Slide 11 Dialogue

The fourth step is to conduct a thorough diagnostic study to identify the quality and service problems.

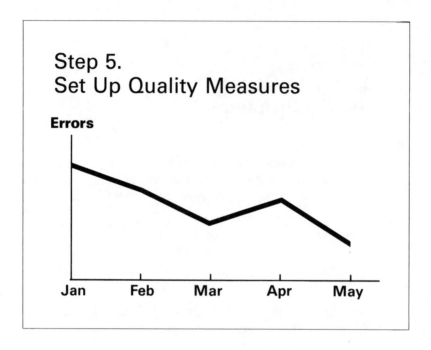

Slide 12 Dialogue

    The fifth step in the program is to identify and define every recurring error, problem, or type of service failure in the department. This process will establish a baseline to document the effectiveness of the corrective strategies implemented by the QA teams. Questionnaire surveys are used to measure the perceptions that clients have of our services.

Step 6.
Measure the Cost of
Correcting Quality Problems

Type of Errors	Number of Errors	Time To Correct	Cost To Correct
Data entry	4	10 mins.	$100
Wrong A/C	8	7 mins.	150
Misrouting	3	15 mins.	90
Total Cost of Errors			$340

Slide 13 Dialogue

The sixth step is to calculate the cost of correcting errors and redoing work. Costs include personnel costs, overhead expenses, and computer costs when there are any. For errors that affect clients, the cost to the client is estimated.

Exhibit 3.14: Slide 14

Slide 14 Dialogue

The last step is to set up awards to give recognition to people who have made an outstanding contribution to the program.

Exhibit 3.15

Proposal to Implement a QA&SR Program

A. Objectives of the QA&SR Program

1. Establish a competent QA&SR program in each operating department.

2. Eliminate conditions that keep work from being done right the first time.

3. Eliminate the costs of correcting and redoing work.

4. Make our organization the standard for quality in the industry.

B. Components of the Quality Assurance Program

1. *Quality Control*

    This component consists of establishing controls to make sure that the services provided to our clients meet all prescribed quality standards. Where there are no quality standards, they must be developed. Quality standards are used to measure the reliability of our services in terms of accuracy, timeliness, and system up time. Quality control also means monitoring clients to keep informed on their perceptions of the quality of our services.

2. *Quality Engineering*

    This component consists of determining the inspection procedures and frequency of in-

(*continued*)

spections. Quality engineering also consists of recording the inspection results, analyzing the data, interpreting the data, and presenting the data to management and to operating personnel so that everyone understands what is being measured and how.

Data pertaining to operative personnel performance should be explained to operative personnel so they know they are being measured and can relate to the performance measures. Operative personnel should not mind being measured if they feel the measures are fair, open, and accurate—especially if the results are favorable.

Inspection data in the form of progress charts and reports will be used to convey quality performance data to operative personnel and working-level management. This shows working-level management which employees are having problems and need additional help and supervision and which employees are doing well. Progress charts also provide senior management with trend data to alert them to the need for corrective action to prevent chronically recurring problems.

3. *Quality Education*

This component consists of the following activities:

    a. orienting people to the concepts and procedures of the QA&SR program and the problems the program is intended to correct

    b. training operative personnel in specific skills, such as telephone handling, processing procedures, routing, etc.

(continued)

c. continually conditioning operative employees and working-level management to work at meeting quality standards.

The payoff of educating people about quality is that you gain their support for the QA&SR program. Additionally, the process of focusing attention on quality and service reliability through participation builds team work.

4. *Quality Management of Input*

This component consists of identifying, investigating, resolving, and preventing quality problems in the work sent in by customers (or other departments) to be processed. The activities grouped under this component are as follows:

a. Inspecting incoming work from clients to make sure it meets standards that we and they have agreed to. This means that we and our clients must reach an agreement on two points: the quality level of the work they send us; and the inspection procedures (sampling plan) we will use to assure that their work meets the agreed quality level.

b. Setting up effective communication networks with key contact people to call when there are problems with incoming work. We must also make sure our clients have an up-to-date list of our people to call when they have questions or problems.

(*continued*)

C. Management Commitment

The implementation process is summarized below. First, executive management must be convinced of the need for a new approach to quality improvement that emphasizes preventing service problems through the use of quality assurance techniques. It is also necessary to present the need for a quality assurance policy that states that everyone is expected to perform in accordance with the quality requirements of the services offered to our clients. Additionally, the need for management to be personally committed to participating in the QA&SR program must be presented. This commitment is required to raise the visibility of the program and to ensure cooperation.

Seven Steps in Building the Program

Steps 1 and 2: Set Up QA Teams and a QA Council

Bring together representatives of each division to form quality assurance teams (QA teams) in each division. The representatives should be able to commit their divisions to action. The QA team should be made up of the division officer, who acts as the team leader, and the managers. The QA project leader should be an active member of all division QA teams. The team members should be oriented to the purpose of the QA&SR program and their role—which is to initiate the necessary action to improve quality in their divisions. The work of the QA teams is part of managing a division, so time should be allocated out of the normal work day for team members to work on team assignments. Executive management should select a QA&SR program project leader who understands and is committed to quality assurance and service reliability. For the purpose of describing the QA team matrix structure, we will assume that a division consists of three to five functional units and is headed by a manager or junior officer. Divisions are organized into a department headed by a senior officer.

With the aid and support of executive management, the project leader should assemble QA teams of representatives for each division and brief them on the purpose of the QA&SR program. The QA teams then run the program. The responsibilities of the members of the QA team are as follows:

a.   To lay out the entire QA&SR program

b.   To represent their division on the QA team

c.   To represent the QA team to their division

d.   To see to it that the decisions of the QA team are executed in their division

e.   To contribute creatively to the implementation of the QA&SR program

The project leader has the responsibility of creating and directing the program. The division QA teams are responsible for administering their own program.

Form a quality council composed of all QA team leaders, the project leader, and key members of the organization's executive management. The council should meet monthly to communicate with executive management and to determine what is needed to continually upgrade and improve the program. The council acts as a central source of information on the status of the program and ideas for action. The council enables departmental QA team leaders to share their problems, feelings, and experiences with each other. It promotes consistency of attitude and purpose. It also exposes the QA team members to executive management and their points of view on quality. Executive management must understand that their participation is necessary to ensure that the direction of the QA&SR program tracks with their goals and objectives and problem situations. They must also provide the project leader with feedback on how they feel the program is progressing and if it is accomplishing what they are expecting.

The QA teams will develop the components of the QA&SR program and present them to the QA councils and executive management for acceptance and support where necessary. Firm dates will be set for accomplishment of each step of the plan, and progress will be reported to senior departmental management. Establishing QA teams does not add to expenses or create more work—it pulls together and organizes quality assurance activities that are already being done, or should be being done, to ensure they are being done effectively.

Step 3: Gain Commitment to Quality

The purpose of this step is to raise all personnel's concern that services provided to clients conform to the standards for the services.

The project leader should conduct a formal orientation with supervisors, managers, and departmental officers to make sure everyone understands each step of the QA&SR program and can explain it to the operative personnel. The orientation program should include participation by senior management to show their support for the program.

Train supervisors and managers to orient employees to the importance of quality. Use communication materials such as booklets, posters, and promotional campaigns to provide visible evidence of concern for quality improvement and to build positive attitudes toward the program. It is important for everyone involved with the program to understand that improving quality will involve a change in the way quality problems are handled. A new structure—a matrix structure of QA teams—will be developed and established. This requires that the people who will be participating in the program be conditioned to accept and be comfortable with the change. Senior managers must understand their role in unfreezing managers and supervisors from their accustomed ways of handling quality problems to the new approach of working through the QA team matrix structure.

Managers must assure supervisors that the new QA approach is not a reflection on their competency to handle quality problems. Otherwise, supervisors might withhold their support of the QA&SR program.

It is important that communication about the QA&SR program, in the form of promotional material, be kept up continuously to show management's commitment to the program.

As part of quality awareness, regular meetings are held between supervisors and operative personnel to discuss specific nonconformance problems and to determine what is needed to correct the problems.

Operative personnel and supervisors are asked to describe problems that keep them from performing error-free work. The problems are directed to the QA team for implementation of solutions. This makes operative personnel and supervisors feel their problems will be heard and answered.

Employees find it difficult to communicate error-causing problems to management and will put up with them because they do not think management cares to correct the problems. Suggestion programs help to an extent, but they are limited; for such programs to work well, employees must know the cause of the problems, know how to solve the problems, and be able to communicate all that. That is a lot to expect of many employees. In order to help operative employees participate, simple one-page forms are supplied to all supervisors to be used to report problems to the QA team. A few simple rules are followed:

1. Every employee and supervisor who reports a problem gets a personal thank-you note. The form is sent to the officer responsible for the department. An acknowledgment memo is sent to the supervisor and employee explaining what action was taken on the problem.

2. If no action is taken, the reasons are given in the acknowledgment memo. If the problem requires a lengthy analysis, this should be communicated so the employee and supervisor know the problem and recommendation are under study.

Meetings of supervisors and operative personnel should be short, positive, and to the point. They should take place on a regular basis. Promises made in these meetings must be kept.

Step 4: Diagnose Quality and Service Problems

A critical part of the QA&SR program is implementing the changes recommended by the QA teams. As causes of errors and service failures are found, ways to eliminate these causes will have to be implemented. Thus, there will be many recurring changes as the QA teams move to improve their operations. To some extent, this means that operations will be in a state of flux during the early stages of the QA&SR program.

Management must also realize that middle management will be investing time in QA team meetings and related

activities. The improvements in quality and service reliability will soon demonstrate that the investment is worth
the time and effort.

The QA&SR program requires support from the training
department, including the senior officer in charge of this
function. It is also important to make sure this officer does
not feel that the QA&SR program will diminish the influence of his training function.

Step 5: Set Up Quality Measures

Quality measures are used to put together the monitoring
system that provides information on the state of quality in
a department and the effectiveness of remedial steps taken
by the teams to correct quality problems. Once the QA
teams are formed and the members understand what they
are supposed to do, they immediately move to determine
what is currently being measured and reported to management. This step gives them an idea of the effectiveness of
the operation.

Departments almost always have some way to measure
errors, backlog, turn-around time, complaints, system
downtime, etc. These measures must be evaluated to see if
they provide management with the kind of information that
is needed to know if the department is processing work in
a timely, accurate, reliable, cost-efficient way.

In addition to measures that focus on errors, backlogs,
and turn-around time, measures that tell management what
clients think of our services must also be covered. Client
surveys must be taken to sample the perceptions that
clients—internal and external to the organization—have of
the services they are receiving.

The QA team must be objective in evaluating the current
quality measures and courageous in establishing new measures. The approach used to identify new quality measures
is to have the QA team members brainstorm about what
they think they and their management need to keep track
of in order to know if their shop is running right. Once the
team comes up with about three to five external and five

to seven internal quality measures, these should be presented to senior management. This presentation can take the form of a mock-up report or perhaps a report with test data obtained by estimates or reconstructed from actual data where available. A questionnaire should also be provided, asking questions such as:

- Is this the kind of information that would help you know if your department(s) are providing their services at a level of quality that meets your expectations?

- If this is not what you think you need, then what do you want?

The questionnaire will be followed by an interview to get whatever additional information is needed. By reviewing and evaluating what is presently being measured, by brainstorming with the department managers, and finally by sounding out what senior management wants, the QA team should be able to come up with quality measures that will effectively monitor and report on the level of input quality, internal quality, and client perceptions of quality.

One example of quality measure reporting is a trend chart display showing the running status for errors or problems by working area. These should be posted weekly to give frequent feedback. The trend chart displays should be large enough to be mounted in the work area.

Another example of quality measure reporting is for a divisional quality coordinator to provide a daily listing of quality problems. These are classified by seriousness, cause, and responsibility. This is done to set the stage for corrective action to resolve the problem.

Senior managers should be pleased to know they will be getting a new system of quality measures to keep them informed of the status of quality in their operations.

Step 6: Measure Cost of Quality Problems

The team calculates the cost of quality-related problems. Examples of quality-related problems are personnel time spent correcting errors, overhead costs of personnel correcting errors, and computer services used to correct or redo defective work.

Step 7: Set Up a Recognition Program

A non-financial recognition award program should be established to recognize people who meet their quality improvement goals or perform in an outstanding way toward quality improvement. People appreciate genuine recognition of performance. It is important that recognition is given for achieving specific goals. It is also critical that the other employees know that the awards were deserved. When this recognition is handled properly, employees will know that management needs their help and sincerely appreciates it. The awards given to individuals need not be valuable—what is important is that management and the employees take the awards seriously.

After the program has been fully implemented and the teams are working effectively to solve quality and service problems, the program enters a maintenance phase. New members should be added to the teams, replacing others who are removed periodically to ensure a constant supply of new ideas and enthusiasm.

An activity Gantt chart will help you plan the implementation schedule for your QA&SR program. An example is shown in exhibit 3.16.

Exhibit 3.16: Activity Gantt Chart

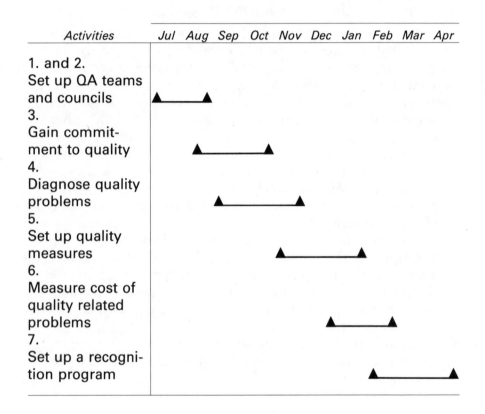

Activities	Jul	Aug	Sep	Oct	Nov	Dec	Jan	Feb	Mar	Apr

1. and 2.
Set up QA teams
and councils

3.
Gain commit-
ment to quality

4.
Diagnose quality
problems

5.
Set up quality
measures

6.
Measure cost of
quality related
problems

7.
Set up a recogni-
tion program

Using an External Consultant

Consultants can be retained to help an organization develop and implement a quality assurance program. Some of the better known consultants in this field are listed below:

> W. Edwards Deming (Washington, D.C.)
>
> Phillip Crosby (Winter Park, Fla.)
>
> Joseph Juran (Wilton, Conn.)
>
> A. V. Feigenbaum (Pittsfield, Mass.)

Most quality consultants will accept a broad range of firms as clients. Very few consultants specialize in service organizations, because the concept of quality assurance is new to such organizations. Until recently, I was one of the few consultants specializing exclusively in this area.

The American Society of Quality Control (ASQC) maintains directories of quality control specialists, some of whom have experience in service organizations. Contact the ASQC for more information.

The advantage of using consultants is that they bring a great deal of varied experience to an organization. Additionally, consultants can interact more freely at all levels of management, because they are not constrained by organizational reporting relationships. Because consultants are perceived as being unbiased, they are able to interact more freely at all levels of management.

Using an Internal Consultant

An organization that is fortunate enough to have on staff a highly regarded, highly motivated individual with excellent communicative, analytic, and interpersonal skills should consider developing this person into an internal QA consultant. He or she can acquire the necessary understanding of the QA&SR approach by attending seminars given by Phillip Crosby & Associates or Joseph Juran & Associates. There are also several books and dozens of articles explaining how to develop and implement a QA&SR program. Most corporate and university libraries provide computer database facilities to produce an up-to-date list of useful books and articles.

With senior management's strong support and visible commitment, the QA&SR project leader pulls together departmental QA teams made up of the departmental officers, managers, and supervisors. Based on knowledge acquired from attending seminars and studying books and articles on QA&SR, the QA&SR project leader explains the QA&SR approach and the responsibilities of the QA team.

Suggestions for developing our custom QA&SR program

Suggestions for developing our custom QA&SR program

Chapter 4

Gaining Commitment for the QA&SR Program

The QA team is responsible for making operative employees and supervisors aware of the importance of building positive employee attitudes toward consistently meeting quality and reliability goals. Positive attitudes are built by conducting presentations to the operative employees, planning special events to focus attention on the drive for quality conformance, and holding meetings to discuss quality-related topics.

Getting the participation of the operative employees starts by inviting them to a coffee and donut meeting sponsored by the QA team. The invitation to the meeting should come from the QA team leader, who is usually the department officer, to show management's support for the QA&SR program.

The best way to invite the operative employees is by sending each one an announcement with a brief description of the QA&SR program. Exhibit 4.1 is a prototype of such an invitation. It is essential to offer an explanation of the program to the operative employees as early as possible to avoid misunderstanding. Also, providing the explanation before the meeting helps the employees to better understand the purpose of the program.

Exhibit 4.1

Invitation to a QA&SR Program Presentation

To: Department Personnel

From: Mr. Greco, Vice President

Subject: Quality Assurance & Service Reliability Program

You are invited to attend a presentation explaining the new quality assurance & service reliability (QA&SR) program. The presentation will be held in the conference room at 8:15 A.M. on January 8. The following is a brief outline description of the program:

- **What is a QA&SR program?**

    A QA&SR program is a planned, systematic way to find and correct the causes of errors. The main idea of the program is to prevent errors from being made.

- **Where will the program be implemented?**

    Initially, two units in the securities department have been selected to pilot the program: The new security issue division and the security redemption division.

- **How will the program work?**

    Quality assurance teams
    Quality assurance teams consisting of the department vice president, the managers, and supervisors will work with Dr. DiPrimio, the QA&SR program

(continued)

project leader, to develop and implement the program in our department.

Commitment

Each of us needs to be committed to preventing errors from occurring. We need to understand the impact that the errors have, and we need to make extra efforts to prevent errors from happening.

Identification of errors

You will be given an opportunity to explain the kinds of errors that are made in your areas. These are the errors we will be looking at initially. As additional errors are identified, we will address them at the QA team meetings. We will look at ways to prevent these errors from happening, whether they are caused internally or externally. That means that we will look at errors we are making ourselves and errors our clients are making on the work they send in to us for processing.

How do we start?

We start by making a conscientious effort to prevent errors from happening.

When do we start?

We have already started. Since the beginning of January, the QA team has met once a week. But the QA&SR program cannot work without your participation. Each of us must make a personal commitment to improving our performance. The QA&SR program will help us toward this goal.

When possible, the QA&SR program presentation should be made at the start of normal working hours. If this is not practical, it should be scheduled to coincide with the operative employees' mid-morning break. The presentation should be held to three-quarters of an hour.

A typical scenario for the presentation is for the QA team leader to begin by giving a brief explanation of the QA&SR program that matches the explanation in the announcement. This talk should emphasize that the program is being implemented because of a company-wide emphasis on quality and service reliability improvement, it is not a reflection on the quality of the work being done by the department.

Management is aware of and appreciates the dedication of the people in the department who are working hard to do their jobs even under difficult conditions. The program is intended to help everyone by eliminating the work environment factors that contribute to quality problems. Finally, the team leader should stress that everyone's participation and commitment to the program is essential to its success. The overall thrust of the QA team leader's comments should be to present the program in a nonthreatening manner to the operative employees and to gain their support.

The QA team leader should then ask one other key QA team member, perhaps a supervisor who is respected by the operative employees, to talk to the group. This person's talk should reinforce the QA team leader's comments and encourage the employees to participate in the program. The supervisor should reinforce the points made earlier by the team leader that the program is not an indication that management is critical of the work being done in the department and that the cooperation and support of the operative employees are essential to making the QA&SR program work. In this talk, the supervisor should also ask the operative employees to think about working conditions (environmental factors) that lead to errors or service failures and to bring these to the attention of their supervisor or the QA&SR project leader. Operative employees should be encouraged to discuss with their supervisors or QA team members ideas and recommendations they have for correcting or preventing the environmental factors that lead to quality problems or to put their recommendations on the Error Cause Removal Form. (See exhibit 4.2 for a prototype of the form.)

Coming from a trusted supervisor, these comments should help ensure the participation of the operative employees.

Exhibit 4.2

QA&SR Program Error Cause Removal Form

1. Description of error or service failure

_____
_____
_____
_____

2. What causes the error or service failure to happen?

_____
_____
_____
_____

3. How does the error or service failure affect the processing of work?

_____
_____
_____
_____

4. Suggestions or ideas for correcting the problem (optional):

_____
_____
_____
_____

_____   _____ _____
Employee   *Phone extension*   *Date*

The last person to talk to the group should be the program project leader. He or she should

> Explain why the organization's management decided to implement the QA&SR program. It is important for the operative employees to understand the reasons that necessitate the program. For example, such reasons may include the need to compete successfully with other organizations that have similar QA&SR programs or to respond to client complaints or criticisms. If employees understand that the need to improve quality and service reliability exists across the entire organization, they will not feel their department is being singled out. They will also feel they are part of an organization-wide effort to improve quality.

> The project leader should also elaborate on the environmental factors that lead to errors and service failures. Examples of such factors should be given in an effort to stimulate the employees to recognize similar factors in their working environment.

The leader should try to speculate as to what might be some of the environmental factors. For example, in departments where lots of data are entered on CRT terminals, errors might be caused by the following factors:

- Documents (forms) from which it is difficult to glean the data to be entered into the terminal

- Data entry protocol, procedures, or screens formats that are not fast and easy to use

- Excessive work loads or overtime that strains endurance

- Lack of verification safeguards built into the software program

The QA&SR project leader should also mention that he or she will be on the work floor studying the work process, the layout of the work flow, and the work procedures.

Another final point to be made at the presentation is that there will be monthly meetings with the operative employees to keep them briefed on the progress being made by the QA team, the disposition of error causes identified by the employees, and planned activities. Additionally, employees will be told of activities that affect work procedures or work flows. And finally, they will be given the opportunity to help the team by providing input and feedback on the steps taken by the team to improve operations.

After the project leader makes these comments, the meeting should be thrown open to questions and comments from the operative employees. The QA team leader and the others should encourage the operative employees to feel free to identify environmental factors that lead to quality problems and to offer suggestions. Employees are rarely at a loss to talk about what makes their job harder to do or causes them to make errors and what management can do to "solve" the problem. Some of what the employees say will not be useful, but intermixed with the trivial, impractical, and useless comments will be those that help identify the underlying causes of quality problems and that provide clues as to how to correct or eliminate these causes. It is the project leader's ultimate responsibility to see that what is significant is separated from what is not. It is also his or her responsibility to see that useful information is recorded in the minutes of the meeting for follow-through action by the QA team at subsequent meetings.

At the end of the meeting, the operative employees should be told that their ideas, suggestions, and recommendations will be carefully considered and that they will be advised of what the QA team decides to do. It is critically important that the operative employees believe that their ideas and suggestions will be taken seriously and examined closely. Otherwise, these employees will not be committed to the program.

Exhibit 4.3 is a prototype of a form used to document the investigation of error causes by the QA team and what action the team took. A copy of the completed form should be sent to operative employees who bring error causes to the attention of the QA team, so they know what was done with their input.

Exhibit 4.3

Error Cause Removal Documentation

ERROR DEFINITION

ERROR CAUSE

FREQUENCY/IMPACT

COST IMPACT

CORRECTIVE STRATEGY

CORRECTIVE PRIORITY

RESULTS

Observing Employees

Observing employees not only gives the QA&SR project leader a first-hand view of work activities and quality problems but also gives him or her a chance to get to know the employees. By talking to the employees about their work-related problems and seeing the environmental factors that lead to quality problems, the QA&SR project leader has the opportunity to gain the cooperation and support of employees.

For example, while observing operative employees sorting packages into canvas sacks, a project leader listened to their complaints that a lot of time was wasted walking back and forth along the long tables, putting the packages into the sacks that were on top of the tables. The employees said sorting would be faster and there would be fewer errors if there was a better way of organizing the sacks. The QA&SR project leader studied the package-shipping operation and noted that the sacks were lined up in alphabetical order: Alabama, Alaska, Arizona, through to West Virginia, Wisconsin, and Wyoming. Some sacks were full to overflowing, some were half empty, and some were empty.

One of the employees, who had worked for the U. S. Postal Service, suggested that instead of keeping the sacks on the tables in alphabetical order, it would be more efficient to hang them on special mail sack racks like the ones in the post office. The racks should be arranged in order of most use: sacks for the locations such as New York, Chicago, Boston, Atlanta that got the most mail should be positioned within easy reach of the mail sorting table, and sacks for the locations that got the least mail should be placed farther away.

At the suggestion of the former postal employee, the project leader visited the post office and saw the racks used to hang empty mail sacks into which mail was sorted. He later persuaded management to purchase similar equipment and arranged the mail racks according to volume. This allowed the employees to drop the packages of mail into the right sack with a minimum of wasted steps.

If this sounds like a lot of cost and effort to solve a minor problem, consider this: thousands of packages of checks and documents are sorted into mail sacks for delivery to hundreds of banks throughout the country. Checks or documents sent to the wrong bank mean that a bank does not get its checks or documents on time and maybe not at all. This causes significant problems to the receiving and sending banks. Getting the mail sorted faster also minimizes the risk of missing a shipment deadline or putting the mail in the wrong sack because of being rushed to meet a deadline. The use of special mail sack racks and the new arrangement of the sacks eliminated a working condition (environmental factor) that contributed to errors.

When the project leader asked the clerk why he had not mentioned the recommendation to the supervisor the clerk laughed. He said, "I told him about it. He said this wasn't the post office, and there was no money in the budget for new shipping equipment. After that I kept my mouth shut."

As a result of installing the new mail sack racks, the project leader gained the support of the shipping clerks, who let the rest of the department's operative employees know that if they wanted something done they should tell the project leader about it. After several more similar experiences, the operative employees and supervisors—who honestly believed they were powerless to get additional equipment—got behind the QA&SR program and really did their best to help the project leader.

Often observing or working directly with employees is the only way to reach them if they have been alienated from management. This is particularly important if the program was preceded by a management-sponsored program, such as quality circles, that failed. Getting the support of employees who have been turned off by an ineffective quality circles program requires showing them how the QA&SR program is different and better, and how it will be beneficial to them as well as to management.

Employee Goal Setting

There may be some advantages to participative goal setting as a way to gain employee commitment, but it is not likely that employees will find goal setting much to their liking. In fact, asking employees to set quality goals is generally pointless. Employees do not deliberately set out to make errors, nor are most errors the result of lack of interest or poor attitude. Most errors are made by conscientious employees who are rushed, tired, inadequately trained, or forced to perform a task under conditions that make it difficult if not impossible to do error-free work. Indeed, asking employees to set error-free goals is also demeaning to the employees and a reflection of management's lack of understanding of operations management.

A much better way to control errors is through statistical analyses that establish error ranges for varying volume levels. When error rates and service failures exceed the established range, management must investigate the cause and take corrective action. Such action might include counseling the clerks with the higher-than-normal error or service failures or changing the system that is causing the problems.

"But doesn't participative goal setting with operative employees gain their commitment to the program and encourage them to meet the goals they help set?" you may ask. No, it does not. Employees feel manipulated by the process of setting goals because they do not want to make errors. They make errors despite their best effort to avoid them. Unlike machines, people cannot perform tasks without any variation. They are subject to fatigue, lagging attention span from repetitive tasks, impaired concentration when upset or pressured, and many other factors.

Goal setting, participative or unilateral, is effective in motivating an individual or a team only when the attainment of the goal is entirely within the control of the employee or team. In such situations, participative goal setting may work—if the employee wants the reward associated with the goal.

If the goal is not within the employee's control, however, goal setting only makes the employee frustrated. For example, if a clerk is expected to make fewer than three errors per thousand complex transactions, the clerk may feel manipulated because he or she knows that at least three, four, or five errors out of a thousand transactions are the result of factors that he or she cannot entirely control.

The futility of setting error goals for employees was clearly demonstrated by Edwards Deming during a three-day seminar held in Philadelphia in 1984. He asked for a volunteer from the audience. This volunteer was assigned the task of using a paddle-like instrument with ten rows of holes to scoop out only white marbles from a container that held both white and red marbles. Each time the volunteer pushed the paddle into the container, which was marked "Raw Materials," both red and white marbles filled the holes. When another volunteer also failed to fill the scoop with only white marbles, Deming "fired" both of them. He called for more volunteers and repeated the demand to fill the scoop with only white marbles. When these volunteers also failed, he fired them, too, saying they were not motivated. Finally, a volunteer solved the problem by taking all the red marbles out of the container. The point was clearly made that to get error-free work, the causes of errors—the red marbles—must be removed. Management must not try to solve quality problems by exhorting employees to do error-free work when the employees do not control the cause of the errors.

Weekly Briefing Sessions

An excellent way to gain the commitment of operative employees to the QA&SR program is to hold twenty-minute coffee sessions once a week to review quality problems. For example, a large financial institution found that a large number of mistakes were being made by the tellers who processed forms filled out by clients to purchase marketable securities. At the meetings the tellers complained that the forms were complicated and the processing procedures were frequently changed without their being notified.

The corrective strategy consisted of holding a briefing session every Wednesday morning from 8:30 to 8:55, before the tellers opened their windows to the public. At these sessions, the supervisor reviewed quality problems and changes in processing procedures. The session provided training, feedback, and a chance for the tellers, who serviced customers all day, to meet with the supervisor and discuss work-related matters. After the briefing session was started as part of the QA&SR program, the tellers felt supportive toward the program.

Interviewing Employees

A final way to gain the participation and support of employees is to conduct individual or small group interviews to get an in-depth view of the environmental factors that cause or contribute to quality problems. This technique is particularly effective when a department head is defensive and reluctant to be open in identifying quality problems and their underlying causes. If the employees are given the opportunity to bring the underlying causes of quality problems to the attention of the QA&SR project leader, and if these factors are then improved, the employees will generally become supportive of the QA&SR program.

Suggestions for developing our custom QA&SR program

Suggestions for developing our custom QA&SR program

Chapter 5

Diagnosing Quality and Service Problems

In order to prevent quality problems and service failures it is necessary to correctly identify their underlying causes. This requires the thoughtful use of diagnostic studies and analyses. An effective diagnostic approach is to conduct focused interviews. These interviews should be conducted by the QA&SR project leader.

Senior Officer Interviews

The first person to be surveyed should be the senior officer of the department: usually an executive vice president or senior vice president. If it seems appropriate, the first vice president or chief executive officer may be interviewed. The focus of the face-to-face interview should be to get the senior officer's views on the quality and service reliability problems in the operation under analysis. The senior officer should be asked for his or her perception of immediate problems and situations that have the potential to become problems. The senior officer should give his or her understanding of the genesis of the problems, the direct causes, contributing causes, the severity of the problems, and what solutions or remedies have been tried in the past or are being considered for the future.

The interview with the senior officer will add perspective to the remaining diagnostic interviews and will ensure that executive management's concerns will be addressed when error and service failure preventative strategies are implemented. The interview notes should be summarized in a confidential memo sent to the senior officer. This memo will document the information gleaned from the interview and minimize the possibility of misinterpretation or misunderstanding.

Interviews with Managers and Supervisors

The next series of face-to-face interviews should be conducted with the officers, managers, and supervisors of the department under study. An interview guide similar to that shown in exhibit 5.1 should be prepared to provide structure and to ensure that the interviewer does not forget to probe important areas.

The interview guide should focus attention on what the officers, managers, and supervisors perceive as the quality and service problems in the department and the causes of the problems.

Exhibit 5.1: Interview Guide

1. Think back over the past six months. What quality and service interruption problems were there? (List these problems with a detailed description of each problem.)

2. What was the impact of each problem on the department, the organization at large, the customer?

3. How was each problem handled?

4. What steps were taken to prevent the problem from recurring?

5. What conditions, situations and/or environmental factors caused or contributed to problems?

6. What ideas or recommendations do you have for correcting the conditions, situations, and environmental factors that cause or contribute to problems? (List ideas and recommendations in detail. Explain that recognition will be given for all ideas or recommendations that are implemented.)

7. What other ideas or recommendations do you have for improving quality and service reliability?

The sequence of interviews should be from the top down: start with the highest-level managers and then move on to other managers and supervisors. This order makes it possible for the project leader to begin with a general overview and gradually acquire a more detailed understanding of the quality and service reliability problems of the department.

The interview notes should be transcribed, coded, and summarized in a diagnostic report. Coding the transcribed notes permits the information to be entered into a computer storage device and grouped according to the type of problem. For example, the causes and environmental factors associated with each problem can be grouped together for study.

Interviews with Employees

If the interviews with senior managers give a view of quality and service reliability problems from the top, the interviews with operative employees give the view from the bottom—or, better yet, from the trenches. In my experience, operative employees, more than any other group, have offered penetrating insights into the situational and environmental factors that cause quality and service reliability problems. If an organization has employees who work on the twilight or night shift, it will be necessary for the QA&SR project leader to come in on those shifts to interview the employees. What he or she learns will make the sacrifice worthwhile.

The same guide used for interviewing managers and supervisors should be used for interviewing operative employees, who often give better-than-expected ideas and recommendations on how to correct the underlying causes of problems. They must be assured that their comments will be treated confidentially and their anonymity protected. Some of the conditions and environmental factors that the project manager should probe for in the face-to-face interviews with operative employees are listed below:

- Excessive overtime, causing fatigue and strain for employees, supervisors, and managers

- Unrealistic time allotments or production deadlines
- Inadequate or ineffective job training
- Improper or inadequate structure and/or organizational design
- Poor or ineffective inter- and intra-departmental communication
- Poor integration of interdepartmental functional relationships
- Inappropriate leadership, management, or supervisory style
- Poor motivation and lack of commitment

While interviewing computer operators on the midnight to 8:00 a.m. shift, I was able to identify the causes of most of the problems that surfaced in the interviews with the supervisor, manager, and senior officer of the department. One problem mentioned by the senior officer was excessive computer system downtime. He was not sure of the cause. There was a high incidence of hardware failures, but he did not think that was the sole cause. He thought that because many programs failed to run to completion (ABEND) the cause was probably software-related.

In the course of interviewing the operators who worked the night shift, I discovered that the cause of most of the program failures was related to neither hardware nor software. Instead, it was a simple failure to communicate processing procedure changes to the third-shift computer operators. Once this failure was known, several corrective strategies were implemented to ensure that the night-shift operators were advised of processing procedure changes. With the improved communication, failures were reduced to a third of what they had been previously.

Interviews with night shift operators, to whom managers seldom talk, often yield insights into the underlying causes and environmental factors of quality problems and service failures. In another example, packages of computer tapes and accounting and billing information were often sent to the wrong client. The senior officer identified this problem as causing a lot of complaints. In some cases, clients took their business to a competitor. Several causes were suggested by the senior officer: sheer carelessness, poor work attitude, lack of adequate staff, and careless delivery truck drivers.

My interviews with the night-shift operators and my personal observation revealed that most of the computer tapes and documents were delivered to the shipping area at 2:00 a.m. The shipping clerks immediately packaged and labeled the tapes and documents for shipping. I checked all the packages and found them to be 100-percent correct. At 3:00 a.m., the shipping clerks were redeployed to other duties. At 4:00 a.m., more tapes and documents were brought down to the unmanned shipping area. The tapes and documents remained in the shipping area until the shipping personnel returned at 6:00 A.M. and found them. With a great deal of annoyance, the shipping personnel hurriedly packaged and labeled the tapes and documents, rushing to get them to the loading platform where the delivery truck drivers were waiting impatiently.

A fast inspection of the tapes and other documents revealed that several tapes had been mislabeled and would have gone to the wrong client. I also observed drivers tossing packages containing tapes onto the trucks. This explained the high incidence of damaged tapes. Obviously, a few nights of on-site observation and interviews with off-shift operative employees can provide a graphic illustration of the work factors that set the stage for errors.

Analyses of Incoming Work

Analyzing work sent in for processing by clients is an excellent diagnostic tool. Quality and reliability problems experienced by a company that processes work sent in by clients are often caused by errors in the work sent in to be processed. In order for a company to improve the quality of its processing operation, the quality of the work sent in for processing must often be improved. To get clients to improve the quality of the work they send in for processing, you must do two things: first, you must document the poor quality of their work, and, second, you must convince clients that improving their quality will result in better-quality service from your processing company.

Documenting incoming quality problems requires an inspection of all incoming work at the initial entry into the processing system. Often the inspection can be performed by the personnel who prepare the work for processing. The supervisor of the preparation function should instruct the personnel on the kind of errors and quality problems to look for.

For example, a large financial institution that processed securities had their preparation clerks identify all the frequently recurring errors that their clients made in sending in securities and coupons to be processed. The preparation clerks were instructed to make photocopies of the errors and to enter the errors in a log. At the end of the month, the supervisor of the preparation function and the QA&SR project leader reviewed all the photocopies and log entries and prepared a report for each financial institution. This report showed:

> The kinds of errors made,
> The number of errors in the time studied, and
> The impact of the errors on the processing company and, where possible, on the client sending in the work.

Then the QA&SR project leader and an account manager visited the financial institutions with chronic quality problems. They explained that they were trying to improve the quality and reliability of the services they provided for financial institutions and, in very brief terms, described the QA&SR program. They asked if the financial institution would cooperate with them in attempting to improve the quality of the work sent in for processing. All of the financial institutions contacted in this manner agreed to cooperate.

In many cases the managers and officers of the financial institutions were not even aware of the errors or problems in the work sent in to be processed. Once advised of the problems and shown the documentation, they took the necessary steps to prevent the errors and problems from recurring. Even the Internal Revenue Service, when approached by the financial institution that serviced them, agreed to take the necessary steps to correct a situation that was causing an excessive number of errors in checks they sent in for processing.

The key to getting cooperation from clients is an analysis that shows and documents the kind of errors made, the frequency of such errors, and the impact such errors have on the processing company and the sending financial institution.

Exhibit 5.2 is an example of the type of report format that you can use to document recurring errors or quality problems with work sent in for processing. The inclusion of cost data underscores the impact of the errors to the processing company. The cost data can also be used to support a cost experience discount. The technique for calculating the cost data is explained in chapter 6.

When errors result in work being delayed in processing or being returned to the client, a strong case can be made for the client to improve the quality of their work by showing the cost and time that can be saved.

Exhibit 5.2: Identification of Errors and Quality Problems Found in Securities Sent In for Processing by the ABC Company, August 1985

Type of Error	Procedures Necessary To Correct Errors	Employee Time Per Occurrence (in minutes)	Frequency of Occurrence	Personnel Cost per Occurrence	Personnel Costs To Correct Errors (Aug)	Personnel Costs To Correct Errors (YTD)
Depositor submitted coupons or bonds that have been "called" for payment	Credit returning paying agent, charge and return shell to depositor	20	15	$3.84	$57.60	$572.16
Depositor submitted future-due coupons comingled with other past-due coupons	Credit returning paying agent, charge and return shell to depositor	20	9	3.84	34.56	299.52
Depositor's calculated dollar value of shell had a difference of $5.00 or less when verified by the paying agent	Process difference account entry and offsetting entry to paying agent	10	6	1.92	11.52	80.64
Depositor's calculated dollar value of shell had a difference of more than $5.00 when verified by the paying agent	Credit returning paying agent, charge and return shell to depositor	20	1	3.84	3.84	30.72
Depositor comingled coupons from other issues into one shell envelope	Credit returning paying agent, charge and return shell to depositor	20	13	3.84	49.92	288.00

Analyses of Employees' Work

Whenever possible, error counts should be maintained for employees who process work. Exhibit 5.3 shows a format you can use to analyze data entry errors. This type of analysis should show all errors made during a month by an employee.

The format shown in exhibit 5.3 can be used to draw inferences about the causes of routing errors. For example, an analyst could draw the inference that John Doe misrouted securities for the following reasons:

1. He made a mistake in reading a coding number and entered the incorrect number into the terminal keyboard. For example, on May 25th he entered "5140" into the system instead of "5170". An examination of the source document showed that the handwritten "7" looked a little like a "4". John's error caused an accounting debit/credit error, caused the wrong client's account to be charged for the transaction, and caused the securities to be sent to the wrong financial institution. All these are expensive errors to correct, and they are highly visible to clients. Getting the client's clerical personnel to write clearer to avoid this type of error is discussed in the chapter on implementing corrective strategies.

2. The error made on April 26th was a typographical error caused by John hitting the "6" key instead of the "5" key. The other errors made on April 26th and on May 10th were of the same kind. John made six typographical errors during the time period under analysis.

3. Another type of error he made was misreading the name of the financial institution. This is a very frequent error, because many financial institutions have almost identical names. For example, on May 20th, John entered Easton Bank instead of Eastern Bank, and on May 13th he entered 1st Fidelity Pa. Bank instead of Fidelity Trust Bank.

This type of analysis is used by supervisors to help data entry clerks avoid making these errors. To further strengthen the learning process, the clerk must also enter the adjusting entries into the system. The analysis also identifies clerks whose error rates are significantly higher than average. When a clerk has a much higher than average error ratio and does not improve with coaching, he or she should be transferred to a more appropriate task.

Analyses of errors and service failures are useful in identifying patterns that show correlations with working environment factors. For example, service failures often occur in late afternoon, when computer telecommunication network traffic gets heavy or backed up. Additionally, error counts jump when computer systems go down for several minutes or hours and then come up, forcing data entry clerks to enter data hurriedly in order to finish their work before going home.

The format shown in exhibit 5.3 can also be used for a department-wide analysis if all the errors made by all the clerks in the department are listed.

Exhibit 5.3:
Analysis of Data Entry Errors Made by John Doe, April/May 1985

Date	Wrong Routing Code No.	Wrong Financial Institution	Right Financial Institution	Right Routing Code No.
Apr 17, 85	0002	ABC Bank	New Jersey ABC Bank	2435
Apr 24, 85	0002	New York ABC Bank	New Jersey ABC Bank	2435
Apr 24, 85	2160	Elmer Bank	South Bank	2185
May 10, 85	2500	NB Bank	NB & Tr Cty Bank	2530
May 20, 85	3790	Easton Bank	Eastern Bank	7850
May 25, 85	5140	NB & Tr Bank	XYZ Bank	5170
Apr 26, 85	7625	Trust Bank	Tr Bank	7525
Apr 26, 85	2425	Toms Rivr Bank	Nat Bank	2435
May 13, 85	7850	East Bank	United Bank	7860
May 10, 85	2486	Clayton Bank	Glo Bank	2530
Apr 26, 85	3345	Producers Bank	North Bank	7359
May 10, 85	2360	Auth Bank	ABC Bank	0002
May 15, 85	7850	East Bank	United Bank	7860
May 7, 85	6050	XYZ Bank	DBA Bank	6175
May 10, 85	4355	BVA Bank	UA Bank	5247
May 10, 85	4150	Deposit Bank	Common Bank	4140
May 13, 85	2425	Toms Rivr Bank	New Bank	2435
May 13, 85	6060	1st Fidelity of Pa. Bank	Fidelity Trust Bank	6080
May 20, 85	7415	Community Bank	Mine Bank	7445

Analyses of Client Complaints

Another excellent source of information about quality and reliability problems are the complaints made by clients. Your institution should have a central source for clients to call when they have a complaint or problem or when they just need information. A "Help Desk" can act as a central clearing house for client problems or requests for information. The Help Desk attendant should maintain a log of problems or requests for information and their status or disposition. Clients complaining about service should be questioned courteously so that you receive as much information about the problem as possible. The problems entered in the log are especially important, because they are the problems and service failures that shape clients' perception of an organization.

Work Environment and Excessive Overtime

Analyses of errors and service failures often show a strong positive correlation with environmental factors. One of the most frequently observed correlations is that between overtime and errors.

An analysis of a wide range of clerical errors arrayed by time of day showed that errors were normally distributed evenly over the eight-hour work day, with a slight increase from 4:00 p.m. to 5:00 p.m. However, on some days the number of errors made between 4:00 p.m. and closing time increased sharply: further study revealed that these error rates jumped on days when the computer system was down or running slowly. Error rates also jumped when clerical personnel worked overtime. Error rates for overtime hours were in the same range as those for times when clerks were rushing to process work delayed by computer downtime.

Another interesting correlation was found between higher-than-normal error rates and clerks who had worked overtime the previous evening. Overall, error analyses showed high jumps in error ratios during overtime hours and during normal hours for clerks who had worked overtime on the previous evening. Sick leave and lateness also increased following extensive periods of overtime. Management must question whether extensive overtime is cost-effective in the light of resulting high error ratios and high sick leave.

Unrealistic Deadlines

An analysis of errors made in billing statements sent out to clients showed an unusually high number of errors being made repeatedly on the statements of a few clients. An examination of all billing statements over a six-month period showed that 28 companies out of 300 had a much higher number of errors in their statements than did the remaining 272 companies. The QA team studying the problem was perplexed until the names of the high error companies were shown to one of the clerks who prepared the statements. Somewhat embarrassed, he pointed out that the 28 companies were at the end of the list of companies to have their billing statements made up. The clerk pointed out that he and his coworkers were always rushing to complete all the billing statements to meet the last mail collection of the day. In their haste, they inevitably made mistakes on these accounts.

A large securities processing company had a somewhat similar problem. An analysis of misrouting errors over a three-month period showed an erratic pattern of high jumps in misrouting errors spaced out over the six-month period studied. Several hypotheses were tested, but none explained the erratic jumps.

An analyst was assigned to observe the four-member group that mailed out the packages of securities. She noted that the fourth member of the work group was absent on every first and third Thursday of the month. This fourth member of the group, an expediter, attended afternoon task force meetings on those days, leaving the work group one key member short. Because they were short, they had to rush to get all the packages ready for the mail pick-up, and the procedure in which the expediter would check each package before it was sealed to make sure the bonds and coupons were being sent to the right financial institution was skipped. The result: misrouting errors on those days were two and three times higher than when the final check was made.

One of the misrouting errors resulted in the loss of three million dollars worth of negotiable bonds and coupons until an armored car driver located the package on its way back from the wrong institution. That could have been an expensive error. As it was, the two-day delay caused a lot of excitement for the management of both institutions and the armored car carrier—all because a final check procedure was skipped. The lesson: small things, if left unattended, can result in big problems.

Inadequate Training

Various analyses and interviews, especially with department managers, often reveal problems that are caused by inadequate or ineffective training. When there is no formal training program and the only training that new employees get is on-the-job instruction from an experienced worker or supervisor, there is a strong probability that these new employees will be inadequately trained. New employees may be taught to do tasks in the same ineffective way as the experienced employees. Poor work attitudes and behaviors are often passed on to new employees by experienced employees. Personnel who make the same mistakes over and over again usually do so not because they do not care but because they do not know any better.

One way to determine if recurring errors or service failures are being caused by inadequately trained personnel is to hold a group meeting and ask questions about work procedures. In a short time it will be apparent if people are clear on how to handle various aspects of their work or if there is a lot of uncertainty.

For example, an analysis of errors made by tellers in completing forms for customers who wanted to buy or redeem treasury certificates showed a high number of errors and omission of necessary information. A meeting with the tellers and supervisors was called by the manager of the unit. The manager asked a number of detailed questions about how to complete the complex forms. She was astonished at the differences of opinion expressed by the tellers. It was clear that each teller was completing the forms according to a different procedure. Unfortunately, in many cases the tellers were wrong. Even the supervisor was wrong in his understanding of several points.

The problem was that the supervisor and the tellers were attempting to interpret complicated instructions from a manual produced by the U.S. Treasury. The instructions, which were difficult to understand to begin with, had been updated by several letters that referred to sections and subsections of the manual. A lawyer would have found it a challenge to interpret the manual and the letters intended to update the instructions. For the most part, the tellers and supervisors were forced to guess how to complete the forms. As a result, errors were made in completing the forms, which resulted in the forms being returned by the Treasury. This caused delays, lost interest payments, and irate clients. To remedy the problem, the manager had the manual rewritten by a professional writer to make it easy to understand and up to date. Additionally, she held weekly training sessions every Wednesday morning, before the bank opened, for the tellers and the supervisor.

Personnel Misfits

Employees who lack appropriate job skills are a frequent source of errors. In some cases, special analyses are needed to identify these individuals. Data entry clerks who input data directly into an accounting system can be difficult to check up on, because most systems do not identify the individual clerk entering the data.

In one department that employs a large number of data entry clerks (about 25), an error-control technique has been devised. Whenever the error ratio for the department exceeds a set range, say 5 percent, an analysis is made of that month's data, grouping the errors according to the clerks who made them. Clerks who show higher-than-average error ratios are coached by the supervisor, who explains whether the errors have any pattern, such as frequent transpositions. The high-error clerks are monitored closely, and if the errors persist, the clerks are reassigned or terminated. Identifying errors by individual data entry clerks is expensive and time-consuming, but it is necessary to root out clerks who are not suited for data entry work.

Inappropriate Organizational Structure

A subtle source of errors and service failures is inappropriate or inadequate organizational structure. By *organizational structure* I mean the way functional responsibilities are partitioned and grouped into departments and divisions. For example, a large financial services organization had an elite department whose function it was to produce the organization's annual budget and other important documents, such as the monthly director's statement. A problem arose, because in order to produce the budget and the director's statement this elite department, which was staffed with analysts rather than accountants, had to use documents prepared by the accounting department. There was considerable friction between the members of the accounting department who prepared the accounting documents and the analysts who had to interpret the accounting documents. Because the budget report had to be prepared under a very tight time constraint, pressure on both the accounting personnel and the analysts often caused small clashes. Working relations between the accountants and the analysts were strained, especially when the analysts had difficulty interpreting the complex accounting statements. The analysts often had to guess how figures were calculated or run to the accountants continuously to ask them. Errors were frequently made as the analysts rushed to meet deadlines. Often the errors were blamed on the accountants, touching off further flare-ups.

The problem here was that the production of the budget and the director's statement were clearly accounting responsibilities and should not have been separated from the accounting department. A similar problem in the same organization was the removal of the pricing function from the accounting department to the marketing department. Here errors were more difficult to detect, but they were still present, because the pricing structure was not supported by effective cost-accounting practices.

Often organizations partition functions not by logic and sound organizational analyses but according to the influence of the players. The result is always a suboptimal organizational structure that allows environmental factors to work against efficient, error-free working conditions.

Lack of Commitment

It is sometimes difficult to ascertain if errors and poor service reliability are attributable to employee indifference and lack of commitment. The QA&SR project leader must look for subtle clues in the way employees interact with one another, with supervisors, and with people outside the department. Signs of hostility or strain in interpersonal transactions are a clue of trouble. Rudeness or lack of interest in dealing with clients or people outside the department are also signs. The clearest sign is the failure to change a behavior or method of doing a task that has been pointed out as being prone to error.

Once it has been established that errors and service failures are attributable to poor work attitudes and lack of commitment, the next step is to determine why employees are that way. Employee indifference and lack of commitment are most often the result of acute dissatisfaction with salary administration (for example, salary levels and increases) or relationships with supervisors and managers. If the problem is with salaries, the QA&SR project leader should discuss the possibility of a separate study to see if the employees have a legitimate complaint that should be addressed.

If the problem is with the way supervisors or managers relate to employees, the QA&SR project leader must determine if the supervisor's or manager's leadership style is ineffective or inappropriate. For example, the supervisor or manager may have a leadership style that is characterized by close supervision, task orientation, quotas, productivity standards, and highly structured job control. This is the kind of supervisor or manager who controls every movement employees make and feels comfortable only when employees constantly check in and ask permission for just about everything they do. This style may not be truly ineffective, but it may be inappropriate for the people and the work environment.

The other extreme is the collegial supervisor or manager who practices a *laissez-faire* leadership style that provides inadequate direction and structure to employees.

Both styles can cause employees to lack commitment to do their best work. In order to correct the situation, the QA&SR project leader must show the senior departmental officer that employee attitudinal problems are attributable to an inappropriate leadership style. If the organization has a competent training director, this person can be enlisted to work with the supervisor or manager to recognize the problem and work to correct it.

These are just some of the ways the QA&SR project leader and the QA teams can identify the causes of quality and service reliability problems and work to solve them.

Suggestions for developing our custom QA&SR program

Chapter 6

Setting Up a Measuring and Monitoring System

The foundation of the QA&SR program is the monitoring and reporting system used to track the quality and reliability measures. The development of a quality and reliability monitoring system starts after the operative employees are brought on board (see chapter 4) and the causes of quality problems have been identified (see chapter 5). The QA team members are briefed on the importance of the monitoring system and how to develop it. Each QA team starts by working with the supervisors and employees to identify all recurring errors and service failures made by employees in the department.

Identifying Errors

Almost all departments maintain error counts. These counts must be studied by the QA teams to determine if they are worth continuing. Existing error counts should be evaluated by the following criteria:

- Is the error made on a recurring basis?

- Is the error significant—for example, is there an impact to the organization or its customers?

- Is the error visible outside the department? Outside the organization?

- Does the error have the potential to tarnish the organization's quality and reliability image?

- Does management rely on the measure?

If existing measures meet these criteria, they should be incorporated into the quality and reliability monitoring system. If not, the value of investing time in continuing to track the existing measures should be questioned, and, where advisable, these measures should be eliminated.

The QA teams should use the recurring errors that are identified through the interviews conducted with operative employees. These errors should be incorporated into the monitoring system. The QA team will work to find the underlying causes of these errors and to eliminate them. The quality and reliability monitoring system will enable the QA team to track and document the effectiveness of the corrective strategies they implement.

The QA teams should determine if errors in the work sent in by clients to be processed are being measured. Few organizations check for such errors or maintain a record of them. If such records are not being kept, the QA team must work with the operative employees and supervisors to identify all recurring errors found in work sent in by clients. Procedures must be arranged to catch errors in work sent in for processing. Information documenting the kind of error, the client, the date and time, and whatever else may be pertinent must be noted. When possible, a photocopy should be made to document the errors for later contact with the client.

Exhibit 6.1 shows a typical monthly error report prepared by one of the QA teams in the securities processing department. The QA team in this department held a coffee and donut meeting with operative employees to identify all the recurring errors they make (internal errors) and the errors made by clients (external errors). In the discussion more than a dozen recurring internal errors were identified. In a subsequent meeting these errors were studied, and the group found that the dozen or so errors could be grouped into about five internal and five external errors.

In identifying errors, it is best to keep the number of errors to fewer than ten internal and ten external errors by grouping errors that are similar in cause and impact. Errors that occur infrequently and that have an insignificant impact should not be counted.

Once the errors and service failures are identified, they must be clearly defined so that everyone has a common understanding of the error or service failure. Any inspection procedure to catch the errors must then be identified, defined, and timed. The QA team must then determine every step required to correct the error and redo the work.

The procedures necessary to correct the error should be comprehensive and should include any activity required by other departments or by the client. Once the corrective procedure is defined, the team must have an analyst or team member observe the procedure several times and record the time required to complete it, including any subroutines, such as making adjustments, answering complaints, or maintaining complaint logs. The observations and timing should be repeated often enough that the team is confident that it has accurately determined how long it takes, on average, to complete the correction procedure or to redo work. The approach used to determine how long it takes to correct an error is similar to the technique used to set time-measurement standards.

Determining the Cost of Errors

The final step is to determine the cost of the time spent correcting errors, making adjustments, and redoing work. The cost is calculated by multiplying the time by the hourly rate of the personnel performing the work, plus the cost of benefits and overhead. If additional computer time is used, it must be included in the cost.

Start with the midpoint of the salary range of the clerks

doing the work. For example, in exhibit 6.1, $13,930 is the midpoint of the salary range for the clerks in the securities department. The benefit package, according to the human resources department, is 40 percent, which adds $5,572 to the employee labor cost. To this must be added an overhead of 17 percent for space and support services—an additional $2,368. An unrefined but easy way to include this load is to add it to the cost of salaries and benefits. The more rigorous approach requires the overhead to be apportioned on the basis of the amount of space determined by employee equivalents. This method is based on the assumption that once space is dedicated to be used for a correction function, the cost of that space must be budgeted for an entire budget period. For the purposes of establishing the cost of errors, adding the per-person overhead charge to the individual salary and benefit cost is close enough.

Thus, the monthly cost of correcting errors and redoing work is figured as follows: the time required to correct errors and redo work is multiplied by the cost of that person's time plus the overhead cost. That sum is multiplied by the number of corrections or the number of times that work must be done to produce the final figure. A year-to-day column in exhibit 6.1 shows the running cost.

Exhibit 6.1: Monthly Error Report, July

Internal Errors

Errors	Procedures Necessary to Correct Errors	Employee Correcting Time per Occurrence (minutes)	July Occurrence	Personnel Cost per Occurrence*	July Personnel Costs to Correct Errors	YTD Personnel Costs to Correct Errors
Shells/bonds routed (sent) to the incorrect paying agent for payment	Credit returning paying agent, reroute and charge correct paying agent	20	72	$3.84	$276.48	$1,950
Daily settlement is not completed on time, due to out-of-balance condition	Recheck posted totals, re-verify, and settle unit	15	10	2.88	28.80	210
Outgoing shipments are not settled on time, due to out-of-balance condition	Re-verify dollar totals of shells per paying agent	15	10	2.88	28.80	210
Incorrect paying agents are charged for shipments of items	Prepare adjusting entries, debiting the correct paying agent and crediting agent charged in error	10	30	1.92	57.60	403
Incorrect general ledger accounts are debited or credited	Prepare proper journal entry into accounting system and correct internal records	10	25	1.92	48.00	336
				Total	$439.68	$3,109

*Personnel cost is based on midpoint salary of $13,930, plus 40% benefit package and 17% overhead loading. The cost is $0.192 per operator minute.

Exhibit 6.1 continued

External (Client) Errors

Errors	Procedures Necessary to Correct Errors	Employee Correcting Time per Occurrence (minutes)	July Occurrence	Personnel Cost per Occurrence*	July Personnel Costs to Correct Errors	YTD Personnel Costs to Correct Errors
Depositor submitted coupons or bonds that have been "called" for payment	Credit returning paying agent, charge, and return shell to depositor	20	29	$3.84	$111.36	$ 784.00
Depositor submitted future due coupons comingled with other past due coupons	Credit returning paying agent, charge, and return shell to depositor	20	13	3.84	49.92	350.00
Depositor's calculated dollar value of shell had a difference of $500 or less when verified by the paying agent	Process difference account entry and offsetting entry to paying agent	10	17	1.92	32.64	237.00
Depositor's calculated dollar value of shell had a difference of more than $5.00 when verified by the paying agent	Credit returning paying agent, charge, and return shell to depositor	20	18	3.84	69.12	490.00
Depositor comingled coupons from other issues into one shell envelope	Credit returning paying agent, charge, and return shell to depositor	20	10	3.84	38.40	266.00
				Total	$301.44	$2,127.00

*Personnel cost is based on midpoint salary of $13,930, plus 40% benefit package and 17% overhead loading. The cost is $0.192 per operator minute.

Definitions of Quality/Reliability Measures

Often, precise definitions are needed of the way quality and reliability measures are calculated. The following definitions are useful in setting up a QA&SR measurement and monitoring system.

MAN HOURS

The man-hour statistic is intended to show the amount of productive time spent on an activity and is used in unit-productivity calculations. The number of man hours reported should include regular man hours worked by all full-time and part-time employees, minus normal lunch periods, absences, vacations, and sick leave plus all *paid* overtime (excludes exempt employees). In addition, outside agency time actually spent on the work of a particular activity (including travel time) should be included. No deductions should be made for rest periods or other idle stand-by time. Regular man hours should also reflect differences in work days between offices or within one office; all employees on incentive plans or other similar arrangements should charge the actual hours they work, not the maximum in the day. The total time reported for each unit should be rounded to whole hours.

AVERAGE NUMBER OF PERSONNEL

The average number of personnel is intended to show the number of employees, officers (except those on leave without pay), and outside agency employees assigned to a particular activity during the report period. This figure is averaged to reflect a result in terms of full-time assignments. Outside agency help includes personnel working on office premises for whom time records are maintained; it does not include off-premises workers who perform services on behalf of the office. The average number of personnel should be obtained by dividing the sum of the *regular* work hours reported in a given activity during the period by the number of *regular* hours one full-time employee could have worked during the report period. If the resulting figure is not a whole number, the fraction should be expressed as a two-place decimal.

The number of hours used in the calculation should be confined to those hours for which an employee, officer, or outside agency worker is paid on a regular basis and should not include overtime hours. No deductions should be made for lunch periods, vacation, sick leave, or other time off for which an employee, officer, or outside agency worker is paid. Payments of vacation pay for individuals leaving the employ of the company should be handled similarly to all other vacation pay. However, the effect of a variable work day schedule, if any, should not be excluded from the calculation.

> Note: Man hours cannot be translated into average
> number of personnel, because the man hour
> statistic is limited to production time.

The following is an example of calculating the average number of personnel:

1. Assume that a new activity began on July 1 and that there was a total of 62 eight-hour days, or 496 regular work hours, during the period July 1 to September 30.

2. Assume also that during this report period three people were assigned 4 hours a day to a given activity, making a total of 12 regular hours daily charged to this activity.

3. Under these circumstances, the average number of personnel charged to the activity for the quarter would be 1.50 $(12 \times 62 = 744 \div 496 = 1.50)$.

4. Assume further that these same assignments were maintained during the period October 1 to December 31, that there were also 62 regular eight-hour work days during the fourth quarter, and that there were 248 such days, or 1,984 hours, during the year.

5. Under these circumstances, the average number of personnel charged to this unit in the annual report would be .75 (12 hours per day $\times$ 124 days = 1,488 $\div$ 1,984 hours possible = .75).

It should be noted from the above example that year-to-date or annual totals of average personnel do not represent summations of quarterly totals but are separate calculations in themselves.

COST PER UNIT OF VOLUME

The cost per unit of volume is calculated by dividing activity production costs by the volume reported for the activity in question. Activity production costs are defined as all costs charged to an activity.

UNIT OF VOLUME PER MAN HOUR

The unit of volume per man hour is calculated by dividing the volume reported for a given activity by the number of man hours also reported for that activity. This statistic can have two variations: hundreds of units of volume per man hour; and thousands of units of volume per man hour.

When activity volume and volume per man hour are both required to be reported at the same rounding level (for example, hundreds of units per man hour and volume reported in hundreds), the calculation indicated above applies. In those cases in which activity volume is reported at one level and the volume per man hour statistic requires another level (for example, units per man hour is required and units are reported in hundreds), care should be taken to modify the calculation noted above so that the statistic is reported correctly.

MAN HOURS PER UNIT OF VOLUME

Man hours per unit of volume is calculated by dividing the number of man hours reported for an activity by the volume also reported for that activity. In all cases in which man hours per unit of volume appears, volume is reported only in units, so that the above calculation always applies.

Examples of Quality Measures

Every department must have quality measures to monitor its processing activities. The following sections provide examples of quality measures that can be adopted for your QA&SR measurement monitoring system.

Errors per 1,000 Transactions

Number of internal errors divided by the number of transactions or items processed (checks, securities claims, endorsements, policies, licenses).

Number of Internal Errors (Types of Errors)

1. Number of transactions with internal errors involving overpayments or underpayments that require subsequent adjustments. Internal errors involve incorrect calculations affecting notices to the client, such as incorrect statements, notices, warnings, and charges sent to clients or parties.

2. Number of errors reported by others (inside or outside the institution) on currency, including: overages, shortages, missorts, missed counterfeits, and missed raised notes.

3. Number of reports or files—for example, Automatic Clearing House magnetic tapes not processed in the processing cycle normally scheduled.

Percent Dollar Holdover Float

Dollar amount of holdover float (funds credited to others with no corresponding setoff debit) divided by dollar amount of transactions, items, or securities processed.

Dollar Amount Difference Account Write-Off per Dollar Amount Processed

The dollar amount of difference account write-offs divided by the dollar amount of transactions, items, or securities processed.

Dollar Amount Suspense Entries over One Month Old per Million Dollars Processed

The dollar amount of entries held in suspense accounts (waiting for disposition) over one month old divided by the dollar value of transactions, items, or securities processed.

Uninvestigated Correspondence Cases over One Month Old

A correspondence case is one requiring internal investigation before an adjustment can be made to a client's account. Each piece of correspondence received and logged in for processing is a case. This measure could be used for any type of case or file assigned for handling, including welfare or regulatory cases.

AVERAGE DAILY BACKLOG

The daily number of unfinished transactions, items, securities, or cases during the month, divided by the number of business days in the month.

NUMBER OF INCORRECT SECURITIES TRANSACTIONS

The number of incorrect definitive (paper document) securities transactions during a month that require correcting adjustments, suspense account entries, or adjustments to clients' balances. Errors that are corrected internally and that do not result in adjustments to clients' accounts or in entries representing uncollectable or unpayable amounts should not be included. This measure includes errors such as, but not limited to:

- Failure to properly switch securities between accounts on a requested date.
- Charging or crediting the wrong client when processing definitive securities transactions.
- Charging or crediting more or less than the appropriate amount when processing definitive securities transactions or account switches.

NUMBER OF BILLING ADJUSTMENTS TO CORRECT FOR
ERRORS PER THOUSAND BILLING ENTRIES PROCESSED

Number of billing adjustments to correct accounting errors divided by the number of billing entries processed.

Most of the above examples of measures were taken from financial and insurance institutions. However, the measures can be easily modified to be used in welfare, housing, governmental, law enforcement, regulatory, or civil service organizations.

Monitoring Client Perceptions

Few organizations consistently monitor their clients to determine their perceptions of service quality and to find out if the services provided are as close as possible to what the clients need and want. Both of these dimensions of quality are "client-driven" measures of quality and they should be monitored.

Questionnaire Survey of Clients

The best way to monitor client perceptions of service quality is through a questionnaire survey. The questionnaire tells management how their organization's clients perceive the quality of the service they receive; the questionnaire survey also serves as a baseline against which to measure subsequent surveys.

A difficult problem with surveys is identifying the exact person to whom the questionnaire should be sent. Even when the client is known, the name of the person who is the primary user of the service may not be known. Sending the questionnaire to a senior officer is not effective, because there is no assurance that he or she will be close enough to the service to know about problems. There is also the risk that a questionnaire completed by a senior officer may reflect information that has filtered up to him or her with all the editing and distortion that occurs as information moves up through an organization's hierarchical structure. One way to avoid this problem is to include the questionnaire in the work sent out to the client with a self-addressed, postage prepaid envelope.

Surveys should also include questions to determine if the services being offered are the ones that best meet the needs of the organization's clients or if additional or different services are needed.

Exhibits 6.2 and 6.3 are examples of a cover letter and a questionnaire you can use to determine what your clients think about the quality of the services provided them.

Exhibit 6.2:
Cover Letter for Questionnaire

Dear Customer:

The check operations department of the High Quality Bank of Philadelphia, as part of its quality assurance program, would like to ask for your help and cooperation in assessing client opinion regarding the quality of the bank's check-shipping service.

May we ask that the supervisor in your check-receiving operation complete all portions of the accompanying questionnaire. If any responses indicate that you are not satisfied with our shipping service, please explain the reason for your dissatisfaction.

We are especially interested in any recommendations or suggestions you can make to correct the cause of any dissatisfaction you may have or to improve our service.

Please return the completed questionnaire in the enclosed self-addressed envelope.

Thank you for your help.

Sincerely,

Exhibit 6.3

High Quality Bank of Philadelphia, Check Operation/Shipping Department, Quality Assurance Survey

This questionnaire should be completed by the supervisor or manager in charge of the check-receiving operation for your bank. Please return the completed questionnaire using the enclosed self-addressed envelope.

1. Are boxes and sealed packages labeled properly?

   Seldom   *Usually*   *Almost Always*
   ☐   ☐   ☐

   Comment:

2. Are you receiving your work according to time schedule?

   Seldom   *Usually*   *Almost Always*
   ☐   ☐   ☐

   Comment:

(continued)

3. Are you receiving your checks packaged properly?

        Seldom   *Usually*   *Almost Always*

        ☐   ☐   ☐

    Comment:

4. Do the bundles in the boxes equal the cash letter to-tals?

        Seldom   *Usually*   *Almost Always*

        ☐   ☐   ☐

    Comment:

5. Are you receiving your accounting statements?

        Seldom   *Usually*   *Almost Always*

        ☐   ☐   ☐

    Comment:

6. Are your accounting statements missing any pages?

        Seldom   *Usually*   *Almost Always*

        ☐   ☐   ☐

    Comment:

(continued)

7. Are you receiving all of your statements and advices?

      Seldom   *Usually*   *Almost Always*

      ☐   ☐   ☐

Comment:

8. What is your perception of the overall quality of our check-delivery services?

      Good   *Marginal*   *Poor*

      ☐   ☐   ☐

Comment:

Bank _____

Name of person completing questionnaire _____

Telephone Number _____

Communicating Findings to Respondents

It is important that the replies to these questionnaires be summarized. The problems that are identified should be given a weight factor that reflects the number of respondents who indicated they experienced the problem.

As a courtesy to the respondents, you should send them a letter thanking them for their participation and sharing the results of the questionnaire.

Some company officers may be reluctant to disclose unfavorable replies about quality, but such a position is indefensible, because trying to hide client dissatisfaction only leads to loss of credibility. It is better to admit to having a quality problem and to emphasize that the company is working to correct it.

Exhibit 6.4 is a prototype of the kind of letter your organization can use to communicate the results of the questionnaire to the respondents.

Exhibit 6.4: Report to Respondents

The check operations department of the High Quality Bank would like to thank you for participating in the questionnaire survey we used to assess customer opinion regarding the quality of our check-shipping service. We would like to share the survey results with you.

Based on more than 200 replies, we learned the following:

  80 percent of the respondents almost always received their boxes and sealed packages labeled properly.

  20 percent of the respondents were not getting their work by the scheduled time.

  90 percent of the respondents usually or almost always received their checks packaged properly.

  70 percent of the respondents almost always found that the number of bundles in the box equaled the cash letter total. The remaining respondents, 30 percent occasionally had a discrepancy.

  15 percent of the respondents indicated that they usually or almost always received another bank's materials mixed in with theirs.

  99 percent of the respondents usually or almost always received their own institution's accounting statement.

  3 percent of the respondents indicated that they frequently received accounting statements with missing pages.

The survey identified several areas in which we must strive to improve our services, as well as areas in which almost all of our clients are satisfied.

(continued)

Through our QA&SR program, we will be working to improve the quality of our check-shipping service; hopefully, we will show more favorable results when we again survey our customers a year from now.

Again, thank you for your participation.

Sincerely,

James Kelley
Vice President, Check Operations

Focus Meetings

Another way to conduct a survey is to question clients on the phone. This may yield acceptable results, but providing answers to a survey questionnaire over the phone is annoying to interviewees.

A better way to determine what clients think about the quality and reliability of your company is to hold focus meetings. These are luncheon meetings to which a small group of clients—five to seven at a time—are invited to meet with key senior officers. These meetings allow management to get the clients' views on the quality of the services they are getting. At the focus meetings, various areas of interest can be probed: the suitability of services in meeting the needs of the clients, new services the clients are interested in getting, and clients' perceptions of the services they are getting. The findings can be used as a baseline against which to measure future focus group meetings.

Monthly QA&SR Reports

An effective QA&SR monitoring and reporting system is essential to track and document the effects of corrective strategies implemented to eliminate the causes of errors and service failures.

From the start of the QA&SR program, monthly and quarterly reports should be produced by the QA&SR project leader and the QA team and distributed to senior management. Appendix A is a prototype of a quarterly report that should be helpful as a guide for drafting reports for your QA&SR program.

The executive summary gives a terse statement of what was accomplished during the period. More detail is provided in the main body of the report. The monthly or quarterly QA&SR report should also include graphic representations of error statistics. The graphs and tables shown with the report in appendix A provide examples that you can modify for your reporting needs.

Suggestions for developing our custom QA&SR program

Suggestions for developing our custom QA&SR program

Chapter 7

Eliminating the Causes of Quality and Service Problems

Once the underlying causes of quality and service problems have been identified through analyses and interviews, the next step is to plan strategies to eliminate these causes.

A frequently encountered cause of quality and service problems is poor communication. For example, a large organization had a centralized computer department that serviced the organization's needs. The computer services department consisted of four functional areas: programming, operations (running the computer systems), technical support, and automation planning. Each functional area was classified as a division and headed by an officer. The operations area was staffed for three shifts.

In each division, there was a QA team made up of the division officer and the managers of the functional units and shifts within the division. The QA&SR project leader, working with the QA teams, identified a host of factors causing quality and service problems. Most of the problems were in one way or another related to poor communication.

There was inadequate communication within the divisions, among the three shifts, among the four divisions, and between the computer services department and the rest of the organization it served.

The poor communication caused friction and even conflict between the computer services department and departments such as accounting, check processing, and marketing, because these departments felt they were not getting the support they needed. The computer services department, on the other hand, felt that the departments they serviced did not understand the constraints under which they had to operate and did not appreciate the service they were getting.

Questionnaire Survey of Service Clients

Many organizations with centralized computer services, processing, and information management departments have problems similar to the ones just described. The cover letter and survey questionnaire format shown in exhibits 7.1 and 7.2 can help determine the extent of these problems. These forms can be modified to serve any centralized service function, such as computer services, loan processing, and central information files for governmental, regulatory, and welfare agencies.

Exhibit 7.1: Cover Letter Accompanying Client Questionnaire

TO: Service Clients

FROM: Computer Services Department

SUBJECT: Client Survey

The computer services department would like to ask for your help and cooperation in assessing client opinion regarding the quality of services provided by our department.

Please complete all portions of the accompanying questionnaire that pertain to your area of responsibility. If any of your responses indicate that you are not satisfied with the service provided by the computer services department, please take the time to explain the reason for your dissatisfaction. Additionally, we are especially interested in any recommendations or suggestions you can make to correct the cause of your dissatisfaction or to improve our service. Please use separate sheets of paper if additional space is needed.

Please return questionnaire to: Computer Services, 4th floor.

Thank you for your help.

Exhibit 7.2: Client Questionnaire

Quality Assurance Survey

Return to: Computer Services
               CSD, Fourth Floor

From: _____ Extension _____

PRODUCTION CONTROL: Responsible for supporting and controlling the daily periodic scheduling and processing of reports.

1. Is production control responsive to your department's needs?

    Seldom　　　　　*Usually*　　　　*Always*

    　1　　　　2　　　　3　　　　4　　　　5

    Please explain and give any suggestions:

2. When an on-line system failure occurs, does production control communicate the status of your system quickly and satisfactorily?

    Seldom　　　　　*Usually*　　　　*Always*

    　1　　　　2　　　　3　　　　4　　　　5

    Please explain and give any suggestions:

(continued)

122

3. Are reports and other work completed according to schedule?

Seldom		Usually		Always
1	2	3	4	5

Please explain and give any suggestions:

4. Is confidential or security data handled to your satisfaction?

Seldom		Usually		Always
1	2	3	4	5

Please explain and give any suggestions:

5. Is quality of printed output and packaging satisfactory?

Seldom		Usually		Always
1	2	3	4	5

Please explain and give any suggestions:

(continued)

NETWORK CONTROL: Responsible for supporting on-line clients by operating the master terminals and providing terminal support service.

6. Is network control responsive and timely in meeting your terminal installation and relocation needs?

      Seldom   *Usually*   *Always*

      1   2   3   4   5

Please explain and give any suggestions:

7. Is network control responsive and timely in meeting your repair and problem-solving needs?

      Seldom   *Usually*   *Always*

      1   2   3   4   5

Please explain and give any suggestions:

8. Is network control responsive in handling sign-on and sign-off problems?

      Seldom   *Usually*   *Always*

      1   2   3   4   5

Please explain and give any suggestions:

In the example above, after the questionnaires were returned, the results were compiled and a memo was distributed to share the results with the respondents. The responses to the questionnaire identified the following complaints:

- Service users did not know who to call when they needed information or help with a service problem

- Service users had difficulty reaching a knowledgeable person to solve a problem

- Service users had difficulty reaching anyone in the service department, because all lines were busy or not being answered when a system was down

- Problems remained unresolved and recurred for long periods of time

- Service users were seldom advised of how long a service or system was expected to be down

Communication Workshop Sessions with Services and System Users

If you suspect that any of your organization's centralized services departments have similar problems, you should begin to improve communications between your services departments and their client users by holding a series of workshop sessions. The communication workshops between the centralized services departments and the line departments they service will give the line officers and managers a clearer understanding of the service departments' organizational structure and problem-management procedures. The sessions will also strengthen communication networks and improve empathy between departments.

These sessions should include the following:

1. An explanation of the way the services departments are structured and of key personnel's responsibilities

2. A description of the typical problem situations encountered by service users and the problem-management procedures that are followed to resolve them

3. An explanation of the ongoing nature of the QA&SR program, emphasizing continuous improvement and periodic surveys of client-user's perceptions of quality

4. A walk-through tour of the service facility or computer area, with a thorough briefing on the computers and all peripheral equipment and controls. Each manager should guide the group through his unit, explaining the equipment and relating the operation to the service it provides to client users. The emphasis should be on what causes the problems that affect the client user and how the service department tries to control these problems and minimize their effect.

5. A question-and-answer period during which client-users will be encouraged to discuss their problems or complaints with the senior officer and managers of the services department. The focus of the dialogue should be on how to reach mutual understanding and empathy between services department personnel and client-users.

The communication workshops should be held in groups no larger than 12 to 15 people. Exhibit 7.3 is a prototype of a letter announcing and describing the sessions.

After the sessions, a questionnaire should be distributed to the attendees to see if they found the sessions helpful and to get other feedback. Exhibit 7.4 is a prototype of such a questionnaire.

Exhibit 7.3: Announcement Letter for Communication Workshop

Office Memorandum

Date: August 8, 1987

To: Internal Service Clients

From: Anthony DiPrimio, QA Program Leader

Subject: Communication Workshop Sessions

This memo explains the objectives, structure, and guidelines for the proposed communication workshop sessions between the processing services department and our internal client departments.

Purposes of the Workshops

To provide a forum for key personnel to communicate about and to jointly resolve problems that affect our customers

To strengthen the working relationship between the processing services department and the internal client departments we service

Who Should Participate

Division heads

Unit Managers

Key personnel, for example, administrators

Management Reporting

Monthly meetings for the first few months

Frequent meetings afterward until no longer productive

Agendas circulated before meetings

(continued)

Minutes distributed to participants and appropriate management people and committees

Action plans

Disciplined follow-up

Structure of Workshop

Shared leadership

Two-hour time commitment per month

Focus on exchange of information and problem identification, resolution, prevention

Bilateral input and feedback

Proposed Agenda

Organizational lines of responsibility and primary contact people

Policies and procedures and how to make them work better

Improving communication and empathy between units

Open forum

Exhibit 7.4: Survey of Participants

To: All Communication Workshop

           Session Participants

From: Services Department

Subject: Briefing Sessions

The members of the services department's QA team would like to thank you for your participation in the briefing sessions. We hope you found the sessions helpful.

In order for us to evaluate the effectiveness of the sessions, we ask you to provide the following information:

1. Did the sessions give you a clearer understanding of the organizational structure of the processing services department and the responsibilities of the managers?

    Yes _____   No _____   Somewhat _____

    Comments:

2. Did the sessions help to give you a clearer understanding of who to call during a service interruption or other related problem:

    Yes _____   No _____   Somewhat _____

    Comments:

(continued)

3. Do you feel the sessions will lead to better communication between you and the services department?

    Yes _____   No _____   Somewhat _____

    Comments:

4. Do you feel the managers in the processing services department now have a better understanding of your operating needs, constraints, and problems?

    Yes _____   No _____   Somewhat _____

    Comments:

5. Were your questions answered to your satisfaction during the sessions?

    Yes _____   No _____   Somewhat _____

    Comments:

(*continued*)

6. Did the sessions give you a clearer understanding of the problem-resolution and decision-making procedures followed by the services department during service or system interruptions?

    Yes _____   No _____   Somewhat _____

    Comments:

7. Do you feel the topics covered during the sessions were treated adequately considering the time available?

    Yes _____   No _____   Somewhat _____

    Comments:

8. In planning future briefing sessions, what topics do you think should be covered? Do you have any recommendations concerning the format of the sessions?

Comments:

Communication Problems within or among Departments

Regardless of whether they are banks, insurance companies, welfare, or governmental agencies, most organizations are structured into large departments made up of smaller divisions. If you survey the officers who manage the divisions in your organization, you will probably find that most of them will admit they have serious interdivisional communication problems. The following are some typical communication-related problems that you might uncover:

1. Poor communication among divisions about project administration details, computer system changes, policy or procedural changes, and contact people;

2. Poor coordination of interface relationships on changes involving projects and systems;

3. Poor definition of shared responsibilities on projects, systems, reports;

4. Unclear definition of boundary interfaces between divisional units;

5. Unclear standards and documentation;

6. Lack of consistency in applying policies and standards;

7. Excessive redundancy in project planning, development, and supervision;

8. Ineffective information flow on projects, systems, policies, and procedural changes;

9. Inadequate involvement of support areas in project planning, causing overcommitment;

10. Unilateral policy decisions by one division that impose burdensome operating limitation on the activities of other divisions.

These communication problems are typical of those found in centralized services departments and the departments they serve. For example, a large commercial bank was organized, as many large banks are, with a centralized processing department. The processing department was responsible for serving the bank's trust administrators, investment portfolio managers, financial markets traders, all the branch banks, and the centralized commercial lending department. In essence, the processing department was the "back-room operation" of the bank.

The processing department had serious problems in coordinating its functional responsibilities with the functional responsibilities of the line departments it had to support. Difficulties occurred with slow turn-around time, large backlogs of unprocessed work, inadequate controls over workflow, and marginal service reliability in some product lines. To solve these problems, the QA teams in the processing department worked to identify the underlying causes: lack of communication, lack of understanding of how the processing department functioned and its requirements, and a poor self-image, which was a carry-over from when processing personnel were treated as just "back-room paper pushers."

Most large departments or agencies that are linked by interdependent functional responsibilities have similar problems, and the underlying causes of these problems are also similar. This is true regardless of whether the department is a centralized computer department, a centralized processing department, or a large administrative agency in the public sector.

Briefing Sessions among Divisions

The best way to correct communication and empathy problems among divisions is to hold a series of briefing sessions among the managers of the functional divisions. These will probably be the members of the QA teams for each of the divisions. A good way to conduct the workshops is to have the officer in charge of the division start off with an overview of his or her functional responsibility. This should be followed by a more detailed description by each of the managers/supervisors of the major subunits within the division. When necessary, disputes between divisions regarding boundaries and overlapping areas of responsibility should be resolved. Unresolved disputes should be set aside to be worked out at later meetings or to be resolved by upper management. To some extent, these interdivisional briefing sessions are similar to the communication workshops held between the services departments and their client line departments. The major difference is that the briefing sessions are held for the benefit of improving communications among the divisions in one department.

Prior to the briefing session, the division officers and managers of the divisions should draft statements of their function responsibilities and distribute these to the attendees. After the presentations of functional responsibilities and related discussions, the group should start to work on problems that were identified in preliminary meetings before the sessions. It is doubtful that the group will get beyond the statement of functional responsibilities and one or two problems before lunch. That is perfectly all right, because the first meeting simply sets the stage for subsequent meetings.

Educational Outreach Programs

An effective strategy for correcting quality problems with work coming in for processing is the educational outreach program mentioned in earlier chapters. There are three steps to this program:

- Identifying clients who have chronic quality problems with the work they send in

- Educating these clients on how to correct their quality problems

- Keeping track of client progress and providing feedback

The first step of an education outreach program requires inspecting incoming work from clients or from upstream departments in the organization. Often employees who process incoming work can already identify the people who have chronic problems. A list should be made of the clients or upstream departments with known problems. Comments should be documented with analyses, samples, and examples of problem work. Documentation should consist of photocopies of the problematic work and the time, date, and recipient of the work. Actual samples of inaccurate work are effective in getting clients or upstream departments to realize there is a problem that must be addressed.

Once the clients or upstream departments with chronic problems have been identified and the problems documented, the next step is to contact the appropriate officer or manager responsible for the work sent in for processing. Phone contact is best, because it is faster and more personal than memos or letters. The client should be told in a pleasant way that the company has implemented a QA&SR program and is trying to improve the quality of work sent in for processing. A member of the QA team should meet with the client or upstream department to discuss problems with the quality of the work sent in for processing. When appropriate, the client should be invited to your place of business to be shown through the processing operation. This is very effective because the client can then see the problems caused by errors in incoming work.

The third step of an education outreach program is to continually monitor incoming work to see if the quality is improving. Monitoring incoming work and giving feedback to clients helps to condition them to send in error-free work. Employee turnover will make it necessary to hold follow-up meetings.

In some cases, work sent in for processing may come from regions that are some distance away. In this situation, it is necessary to hold workshop meetings at other locations. For example, if your organization receives work for processing from clients scattered in outlying regions of the state, hold a series of workshops in those locations at hotels or inns that offer suitable meeting facilities. Discussions should stress how to send in work that is error-free and the benefits to the client of error-free submissions: fewer returns, fewer adjustments, faster turn-around time, etc. A side benefit of the workshops is that they build goodwill and demonstrate your organization's serious commitment to quality. A small block of time may be given to the marketing department to market a new service—an added benefit for your organization.

Educational Material

Often a QA team will find that the cause of problems with incoming work is simply that clients do not know the correct way to prepare work to be sent in for processing. For example, one organization found that almost half the work sent in by one client for processing was incorrectly prepared. The organization that processed the work routinely corrected the errors and filled in missing information. Using a work measurement technique, the QA team found that the time spent correcting and filling in missing information on incoming work amounted to the equivalent of 2.5 clerks. When these costs were revealed to managers, they were surprised and dismayed, because the price set for processing the work never accounted for correcting errors and filling in missing information. The price was set based on processing error-free work from completely filled out forms. The organization had been subsidizing their clients' processing operation by 2.5 clerks.

Further analyses showed that missing information usually required the client's clerks to look up reference numbers. For years these clerks either guessed at the reference numbers or ignored them. Why should they bother to look them up?—the organization that processed the work would take care of that. And indeed, the organization had been doing their work for years.

One of the first things a QA team does is to check the quality of the work coming in for processing to see if it is error-free. If it is not, the QA team sets up strategies to control the quality of incoming work. In this case, the QA team decided that because of the large number of clients submitting work for processing, they would prepare educational material showing their clients the correct way to send in the forms for processing. The QA team identified 200 clients with the highest volume and the most errors. Special educational material was developed explaining how to correctly fill out the complex forms, how to calculate the various totals, and the importance of filling out the forms correctly and completely. The educational material also included a statement that incorrectly completed forms would be returned unprocessed.

The educational material was sent to the officer responsible for the operation, with a letter that politely explained the need to correct the problem. The letter made it clear that the company could not continue to spend expensive clerical time correcting forms without raising the cost of the service.

The educational material, the request for cooperation, and follow-up calls by the unit supervisor worked astonishingly well. Errors and missing information on incoming forms were reduced by 89 percent in the test group of 200 clients. Many clients were surprised to learn that their clerks were not filling out their forms correctly. Indeed, the client could not be expected to know that work sent in for processing was causing problems for the processing company. It was up to the processing company to bring the problem to their attention, as this company did through its QA&SR program.

How much of your employees' time is spent doing work that should be done by your clients? If your organization is like the one in the example, this amount is probably significant, perhaps as high as 20 percent of the total time spent in preparing incoming work prior to processing.

Reviewing and Updating Forms and Manuals

Forms that are difficult to understand are a frequent cause of errors and omissions in incoming work. An analysis performed by a financial organization of incoming forms completed by their clients found that certain parts of the form were frequently filled out incorrectly or omitted because people filling out the forms were not sure what to put down. The QA team completely redesigned the form and client-tested it. After several more redesigns, an analysis of client-completed forms showed that the number of errors and omissions was reduced by 70 percent.

To further reduce errors, instructional signs that showed how to fill out the forms correctly were provided by the processing company. These signs were posted in the area where the forms were completed by the client's clerks.

When clients or employees depend on manuals for instruction on how to prepare work, these manuals should be analyzed by the QA team to determine if they are clear, definitive, and easy to understand. If not, they should be rewritten by a professional writer, or at least someone with excellent writing skills. Many organizations rely on manuals that cannot be understood by employees with marginal reading comprehension skills. Professional technical writing skill may be necessary to produce manuals with illustrations that are within the reading comprehension level of the people who must use them. The reduction in errors should offset the cost of producing the manuals. A manual that is not used effectively is worse than worthless—it generates errors.

Standardized Preprinted Control Documents

Another frequent source of input errors is the wide variety of source documents clerks must use to enter information into a CRT terminal or another source document. An analysis performed by a QA team in one financial services organization found that clerks had to hunt through a large volume of widely varying application forms to find the information they needed. Hunting to find information consumed time and caused a high rate of input errors.

A control group using a wide variety of application forms from different clients was tested against an experimental group using forms with just one format. The experimental group using a consistent format had an error rate less than half that of the control group, and the productivity of the experimental group was more than twice that of the control group. Based on these statistics, an analysis was made that showed the cost of supplying clients with well-designed standardized forms was justified.

An example of a useful format for presenting information is a standardized cash letter form for incoming check deposits with preprinted information, such as the client financial institution's ABA number and various control codes. The cost of the preprinted form for individual clients is not high and is outweighed by the lower error rates and increase in productivity.

Software Modifications

The sequence followed by a clerk in entering data into a terminal should track exactly with the sequence in which data appears on the printed form used to key in data. Any variance between the form's data-presentation sequence and the program data-entry sequence is a potential source of data input errors.

Another corrective strategy that should be considered in preventing data input errors is that of modifying or enhancing the computer program to provide computer-produced verification checks. Such system enhancements can be programmed to provide immediate verification and the facility to make corrections when the data entry clerk sees a discrepancy or error. System enhancements are expensive, but they are often justified and should be considered as a means of controlling data entry errors. Wherever possible, weekly or monthly print-outs should be produced showing error listing by individual data entry clerks. These error listings should be used for one-on-one coaching by the supervisor and the clerk. Such coaching is very effective in heightening awareness of the importance of error-free work.

Training

A frequent cause of quality problems is inadequate training. Fortunately, training deficiencies are easy to remedy. The first step is to assess the extent of training deficiencies. Training should be tightly focused on specifically what the operative employees and supervisors need to know.

When possible, video-assisted training should be used. For example, data entry clerks, tellers, and processing operators can effectively use video training programs that *show* the right way to perform a task or activity. Today's clerical employees, more than those of previous generations, are oriented toward "show-me-how" training. Whether management and training specialists like it or not, most entry-level employees spent their formative years watching and learning from television. Management must accept this fact and capitalize on it by providing video-tape training showing employees the right way to do their jobs.

An especially effective way to use video-assisted training is to use a portable television camera to make a tape of an employee performing a task or activity. Then hold a training session at which the tape is shown. Let employees critique the way the employee performed the task on the tape, pointing out what was done correctly and what was done incorrectly. This critique approach will enable whoever is conducting the training session to explain in concrete terms the right and wrong way to perform a task or activity. It is an interesting and enjoyable way to train operative employees. It also helps supervisors play a coaching role, which strengthens their influence over the people in their units.

This chapter presented corrective strategies used successfully by QA teams. The key to successfully using these strategies is to let the choice of the corrective strategy be dictated by the nature of the quality or service problem. Too often the wrong cure is applied—one that is not responsive to the nature of the problem. A careful, penetrating analysis of the quality or service problem is the best way to determine the most direct remedial strategy to use.

Suggestions for developing our custom QA&SR program

Chapter 8

Documenting the Program Benefits

The best way to document the benefits of a QA&SR program is through the minutes of the QA teams, monthly executive summaries, quarterly reports, and end-of-year reports. Every QA team meeting should be preceded by an agenda of items to be covered at the meeting. Minutes should be kept to record all important points discussed at the QA team meetings. Reports, forms, and documents discussed at the meetings should be included in the minutes to ensure complete documentation of the team's activities. In addition, periodic surveys of customer perceptions of service quality should be recorded in the minutes. Exhibit 8.1 is a prototype of an agenda sent out to notify a QA team of a meeting.

Exhibit 8.1: Agenda

To: Computer Services QA Team

From: Anthony DiPrimio, QA Project Leader

Subject: QA Team Meeting Agenda

The next QA team meeting is scheduled for March 3rd at 10:00 a.m. in Conference Room 2.

The QA team is expected to focus on the following agenda items:

1. A review of the issues raised at the workshop session. Pat Mason will present a report summarizing the issues and recommendations for approval by the team.

2. A review for approval of the procedures for scheduling system tests. The proposed procedures will be presented by Bill Willard.

Material to be presented for review by the QA team will be circulated three days before the meeting. Please notify the team leader of any additional agenda items at least three days before the meeting.

Minutes of the QA Teams

If there are several QA teams in a department or division it is best to consolidate the minutes of their meetings. For example, if there are four QA teams functioning in a department, the minutes of all four QA teams' meetings should be consolidated into one report. This allows members of one QA team to be kept informed on what the other QA teams are doing.

Exhibit 8.2 serves as an example of the type of format that can be used to promulgate the minutes of four QA teams from one large department. One of the problems that all four QA teams addressed was inadequate communication among the divisions represented by the QA teams. Distributing copies of the minutes to all QA team members helps to communicate what each QA team is doing.

The workshop session referred to early in the exhibit 8.2 minutes was conducted to address the problem of inadequate communication among the divisions. Because communication problems seem to be common in large departments, the minutes describing the communication workshop were reproduced to provide you with a guide for similar workshops in your organization's QA&SR program.

The minutes provide excellent documentation of what was covered at the workshop and the benefits achieved. Whenever a report to management is needed, the minutes can be reviewed. Monthly and quarterly reports can also be produced using information gleaned from these minutes.

Exhibit 8.2

Minutes of the Computer Services Department QA Teams

Data Services Team
Ron Shell, V.P., Team Leader
Jack Cooper, Mgr. Prod. Control
Ron Harvey, Mgr. Quality Control
Steve Glass, Mgr., Net. Control
Harvey Gorman, Mgr., Oper. Control

Automation Planning Team
Stan Fisher, V.P., Team Leader
JoAnne Smith, Off. Auto. Plan.
Gary Evans, Mgr., Prod.
George Will, Mgr. Fin. Ctrl's.
Alice Sheldon, Mgr. Bus. Sys.

Technical Services Team
Bill Willard, V.P., Team Leader
Bill Houseman, Mgr. Oper. Sys.
Don Edwards, Mgr. Tech. Plan.
Glen Evans, Mgr. Data Services
Pat Mason, Mgr. Communications

Systems Development Team
Ed Flana, V.P. Team Leader
Joe Cross, Mgr. Pay. Sys.
Rick Lock, Mgr. Acc't. Sys.
Marie Brooks, Mgr., Trust
Victor Kellman, Mgr., Fiscal

The technical services and systems development QA teams held their first workshop session.

1. Bill Willard opened the meeting with the following overview of his functional responsibilities:

   Implementation and support of operational and environmental software products
   Support of production systems
   Systems security

2. He then identified and defined the following problem areas:

   Inadequate project planning
   Inadequate involvement of technical services in setting project commitments

(continued)

146

3. Because of these problems, Bill found it difficult to effectively control the scheduling and deployment of his resources. It also made it difficult to plan and meet his department's initiatives. In order to provide for adequate lead time he issued a directive requiring several days' notice to schedule a request for systems support. This has resulted in his department being viewed as not being responsive.

4. Ed Flana gave the following overview of his functional responsibilities:

   > Check and ACH support systems
   > EFT and Fedline systems
   > Trust systems
   > Statistics reporting systems
   > Accounting systems
   > Fiscal systems

   Ed defined the following problems:

   > The need for adequate test facilities and test scheduling
   > The need for better responsiveness
   > The need for more clarity in defining areas of functional responsibility in technical services
   > The need for better communication and coordination
   > The need for bilateral input in policy making that affects his department

5. Following Bill's and Ed's presentations, there was a general discussion of the following problem area:

   > Inadequate test facilities
   > > System users find it difficult to schedule tests because the computer systems are available only on weekends. This delays testing.

(continued)

6. The following recommendations were made:

   Systems development managers will be more precise in specifying their test design needs and in communicating these to technical services.

   Technical services will be more precise in defining the specifications of test facilities and their availability.

   Users will be advised of the problems in matching test design specifications and test facility availability and the resulting risk factor.

   Adequate lead time, consistent with user needs, will be provided to schedule testing.

7. Bill indicated that his decision to require several days' lead time to provide support was necessitated by the need to be able to balance his department's initiatives with the short-term support requests flowing in from other areas.

8. The systems development managers indicated that they are frequently called upon by client-users to respond quickly to their needs. Because they must respond immediately, they are forced to go to Bill with short-lead-time requests for system changes. Bill suggested that part of the problem is that systems development's role in reacting to client-user requests for service is not taken into consideration when technical services plans and schedules the utilization of its resources.

(continued)

9. The following remedial recommendations were made:

A monthly meeting of the officers and managers from both departments will be held to focus on project plans for the upcoming four-week period, any coordination problems, pass-through information, and other information.

The monthly meeting will serve as the forum for addressing issues affecting both departments and will ensure a high level of integration of both departments' functional responsibilities.

10. The workshop ended on a positive note, with the group feeling that the process of providing better integration between the two departments had begun.

11. The next QA team meeting is scheduled for March 12th at 10:00 a.m. and will focus on the procedures for the new problem-management system.

Quarterly Reports to Executive Management

A quarterly progress report should be prepared for senior management, summarizing the QA teams' efforts, progress, and plans. The quarterly report should focus on what senior management wants to know about the progress of the QA&SR program.

The QA&SR project leader should distribute advance copies of the quarterly report for comment to the QA team leaders before releasing copies to senior management. This is more than just a courtesy; it ensures that the contents of the report are accurate. The QA team leaders can spot errors in the report and advise the QA project leader to make corrections before the report is distributed.

The quarterly report should always start with an executive summary that highlights the important points of the report. The executive summary can be very brief, as it is in exhibit 8.3, or more comprehensive, as is the one that precedes the prototype of a quarterly report in appendix A.

Exhibit 8.3: Executive Summary

This report, in broad terms, presents what the QA teams accomplished in the first quarter. Their efforts focused on the following tasks:

- Expanding the QA team structure to encompass all major areas in the securities, checks, and computer operations
- Structuring communication networks to improve integration among divisions and departments
- Improving understanding of divisional and departmental responsibilities
- Maintaining commitment to error-free work through continuous monitoring of error rates
- Identifying financial institutions with quality problems and helping them—through an educational outreach program—to improve the quality of work sent in to the bank for processing.

Additionally, quality improvement targets were set for all the teams, and their full commitment will be continued throughout the year.

Whenever possible, the minutes of the QA teams, the minutes of the QA councils, any special reports or analyses, the quarterly and end-of-year reports, and all statistical exhibits should be maintained in a PC or mainframe database for speed and efficiency in producing reports. Efficient monitoring and documentation of the effects of the QA&SR program is essential to its success.

An end-of-year report should be produced for executive management. This report should set forth what was accomplished during the year, with extensive documentation of improvement in quality, service reliability, and productivity. The end-of-year report draws from the quarterly reports and is similar in format.

Suggestions for developing our custom QA&SR program

Suggestions for developing our custom QA&SR program

Chapter 9

Integrating the QA&SR Program into the Managerial Process

One of the distinguishing characteristics of a QA&SR program, and its major advantage, is that it is administered by a QA team composed of operating officers and managers. The same people who are responsible for running the department also run the department's QA&SR program. This means that as the QA&SR program progresses, its processes become assimilated into the department's managerial processes and functions. After the QA&SR program has been functioning for about a year, the activities of the QA teams are indistinguishable from the general management of the department. Indeed, the QA team meetings take on the appearance of departmental staff meetings. At that point the QA&SR process has been completely integrated into the managerial process.

Incorporating the Program into the Organization's Strategic Business Plan

It is essential that the QA&SR program constitute a major component of the organization's strategic plan. This is critical because the strategic plan sets forth the direction, establishes the goals and objectives, and allocates the resources—personnel, budget funds, computer systems, and facilities—for the organization. Indeed, the organization's ability to realistically plan its marketing strategies depends as much, if not more, on its ability to differentiate its ser-

155

vices based on client perception of superior quality as on any other factor. An example of a goal set by a bank with an effective QA&SR program is to maintain attention to innovation and quality to develop client loyalty, which is essential in developing new service lines.

A QA&SR program raises the quality and reliability of an organization's services to the highest level achievable. Doing so enables the organization to shape client perceptions, causing them to view the services as different from those of competitors and worth a reasonable price difference. A QA&SR program also raises operating efficiency to the achievable maximum through the processes and techniques described in earlier chapters. This keeps the costs of providing services as low as possible. Such cost containment makes it possible to institute selective service pricing, which further enhances the organization's ability to develop marketing strategies.

These are but a few of the more obvious reasons why the QA&SR program must be an important part of the strategic planning process and must figure prominently in your organization's strategic plans.

For example, a typical strategic plan starts with a statement of overall direction that sets forth the organization's basic strategies for achieving goals and measuring organizational performance. Marketing strategies focus on market analyses used to identify market trends, competitor trends, client needs and preferences, and regulatory considerations. Internal analyses focus on the organization's managerial environment, analyzing its perceived strengths and weaknesses. The QA&SR program impacts equally on the external market factors, by influencing client perceptions, and on the internal factors, by improving operational efficiency.

If your organization is willing to make the working assumption that it can influence external market factors, the objectives and goals of the QA&SR program must be incorporated into the strategic plan. One way to incorporate the QA&SR program into your organization's strategic plan is to link business development strategies with tactical quality improvement targets and goals. For example, your organization may be providing a service that could be provided more quickly and reliably as a result of operational improvements brought about by the QA&SR program. The service might even be able to be enhanced to the extent that new service products could spin off. These new service products are the result of the quality assurance process. A new service product may come about as the result of being able to deliver a service earlier in the work day, with faster turn-around time, or in greater detail.

Thus, what may at first appear to be just improved service and reliability can be refined and redefined into new service products. In the banking industry, the ability to get check processing documents and checks to a payor bank before 10:00 a.m. created a new service product called, logically enough, Payor Bank Statement Accounts, which permits more efficient money management. Many new bank services are the result of faster, more efficient processing of checks, related transaction data, and related accounting documents. One such service is the High Dollar Group Sort Service, which is a deposit program that offers accelerated collection of checks drawn on selected banks found outside of cities in which Federal Reserve offices are located. This service provides more efficient check collection by giving depositors better availability for checks drawn on banks in these cities.

Many other possible services could be marketed that are no more than refinements or extensions of more basic services—refinements made possible through improvements in operating efficiencies.

Integrating the Program into the MBO Process

Most service organizations have management by objective (MBO) programs. The QA&SR program should be integrated into the MBO program. This is accomplished by setting quality objectives or goals and incorporating them with the department's other objectives into the MBO program.

Examples of quality objectives are listed below:

- Set up a QA&SR program for the accounting department

- Achieve and maintain an error ratio of less than 2 percent per 100,000 items processed

- Achieve and maintain 100-percent up time for the communication network system

Quality objectives can also be stated in terms of major quality activities, such as these:

- Conduct four educational outreach seminars

- Conduct a questionnaire survey

Quality objectives, like all objectives, should be attainable but challenging and documentable.

Integrating the QA Monitoring System into Other Management Monitoring Systems

All organizations have management monitoring systems that are reviewed periodically by senior management. These monitoring systems typically include trend charts on unit costs, sales, progress on major projects, cost-revenue ratios, etc. The statistics gathered in the quality measurement system on error ratios, client perceptions of quality from questionnaire surveys, cost of quality, and other quality statistics should be incorporated into the management monitoring system that is reviewed at monthly meetings with senior management. Specific examples of how to show quality statistics were provided in earlier chapters.

In addition to incorporating reports on the progress of the QA&SR program into the management monitoring system, it is also essential to include frequent reports of the QA&SR program's progress to your organization's directors, shareholders, and important clients. Exhibit 9.1 is a prototype of the kind of statement that can be included in your organization's annual statement or report to shareholders to describe your QA&SR program.

Exhibit 9.1: Brief for Annual Statement or Shareholders' Report

Quality Assurance

This organization's commitment to quality was further advanced in 1987 with the implementation of a new QA&SR program in several high-activity areas. The QA&SR program is proving that it is better to avoid making errors than to spend time and money later to correct them.

At the beginning of the year, quality assurance teams went to work in several units of our securities department, in which the volume of work and tight deadlines make accuracy critically important. The teams' job was to identify and document the sources of problems and errors—internal and external—and to develop procedures to correct or minimize such problems in the future.

Results were swift and dramatic. In just twelve months, internal errors in the Treasury new issues unit dropped nearly 80 percent in spite of a 54-percent increase in work volume. Similar success was achieved in the municipal bonds division.

In addition, a special educational program was implemented to reduce external errors and to help client banks avoid unnecessary reprocessing. Forms were revised and simplified, a new and easier-to-use Treasury securities guide was published, and special workshops were held for more than 250 of our clients.

A slightly more detailed statement, similar to the prototype shown in exhibit 9.2, should be included in the end-of-year report to the board of directors. The prototype also provides some insights on how a mature QA&SR program makes continually greater contributions to the organization as time passes. The program's role may be expanded in later years to encompass responsibility for ensuring that present and new service lines meet client's needs and expectations.

Exhibit 9.2: Statement for End-of-year Directors' Report

Quality Assurance & Service Reliability Program

The QA&SR program is now in its second year of operation, with groups established in the securities, computer services, check operations, and EFT functional areas. The program will be expanded to the accounting and cash operations areas in 1988. Our personnel have a heightened awareness of the importance of error-free work and high service reliability as a result of the program.

Internal Accomplishments

- Quality measuring and monitoring systems have been established.

- A Network Control Service "Help Desk" was implemented and is functioning effectively.

- A computer-related problem resolution/management system is in operation.

- New processing controls have been adopted to increase efficiency and improve quality and service reliability.

External Accomplishments

- A quality outreach program has been established to visit clients with chronic quality problems, either with work submitted for processing or work received from our organization.

(continued)

- Questionnaire surveys were conducted to ascertain financial institutions' impressions of the quality of our services.

- A series of seminars was conducted to brief clients on new services and on how to avoid errors in preparing work sent to us for processing.

Plans for 1988–1989

In addition to maintaining full commitment to the present QA&SR program, plans are being developed to expand the service reliability dimension of the program. A formal policy on service reliability will be adopted. The policy will establish authority and accountability for service reliability within our organization.

We have already started to develop our position on service reliability by holding focus group meetings with our client base to make sure that we understand their service reliability needs. This is being done to ensure that our services are responsive to their requirements. An organizational structure will be designed to support the expanded commitment to service reliability.

Implementation of the organizational structure is scheduled for first quarter 1988. The QA&SR program will assume responsibility for making sure that present and new service lines offered to clients meet reliability standards.

Integration of the QA&SR program into long-range automation and service line planning is essential to ensure uniform quality and service reliability. The extension of the QA&SR program into automation and service line planning will be driven by the emphasis on new developments in telecommunication inquiry capability, access to data base services, and interstate communication and service links. The expanded authority of the QA&SR program will give our clients a high level of confidence that they can depend on our organization's service lines.

Meshing the QA Program with Incentive Plans

Are quality assurance programs and performance-based compensation programs mutually exclusive or mutually supportive? In order to answer this question, it is necessary to clarify what incentive plans are supposed to accomplish. Well-designed, well-implemented, and well-administered performance-based compensation programs claim the advantage of increasing productivity. In addition, because they emphasize sharing the benefits of increased productivity with managers, supervisors, and especially the operative employees, such programs usually have the support of everyone who participates in the program.

Another claimed advantage is that incentive programs provide employees and management with standards by which to measure performance, thereby reducing uncertainty. Proponents of incentive programs claim that employees like the program because they can use it to document their productivity and quality at performance evaluations.

Of course, the most important claim is that of substantial increases in productivity. One consultant claimed an 11.8-percent gain in productivity within nine months of the installation of an incentive program (William Abernathy, *Bank Notes*, Summer 1985).

The story of the Lincoln Electric Company of Cleveland, Ohio is often cited to support the incentive pay approach. (Reports of the program appear in various published shareholders and directors statements.) The company paid each of its 2,405 employees an average of $17,380.00 in year-end bonuses for a total of $41.8 million in one recent year. Bonuses averaged 97 percent of employees' annual earnings. A pay-out of that magnitude must be substantiated by productivity increases of even greater magnitude. And apparently this was the case.

In general terms (a more in-depth explanation will be provided later in this chapter), the incentive approach sets up productivity standards. Operative employees are paid for what they produce, based on the standards, rather than for their time. Because operative employees' production is monitored by the incentive system, presumably, less supervision is required and wider spans of control are possible, thereby reducing the number of supervisors needed.

Incentive plans have been criticized, however. One criticism is that managers and supervisors may resist implementing and administering incentive programs because they feel threatened by such programs. They may feel that senior management considers that the program is needed because the managers and supervisors did a poor job of motivating their subordinates to meet acceptable productivity levels. Regardless of senior management's protestations to the contrary, the line manager and supervisor may resent the program as an indication of management's lack of confidence in their ability. In addition, they may resent the idea that a consultant's program is needed to ensure adequate utilization of personnel.

These defensive or resentful feelings tend to affect senior management's relationship with line managers, becoming another wedge splitting the already strained relations between line managers and senior management. Because most managers already feel alienated from management, this causes further deterioration of the working relationship.

Another limitation of most incentive programs is that they take for granted that what is being done in a department is what should be done. They may miss important ways to increase the efficient utilization of resources by failing to determine if what is being done should be done at all or if it should be done differently.

A third criticism of incentive systems is that it is hard to establish productivity standards for jobs that require work that is not homogeneous and for jobs that are performed by only a few employees. With small groups of employees, it is difficult to establish productivity standards, because the productivity standards are based on a limited sample. There is little assurance that the standards are set at the right level to ensure adequate utilization of the employees.

The verdict is yet to be reached on whether incentive programs are an effective way to achieve long-term optimal utilization of human resources. What is clear is that a quality assurance program, with its matrix organizational structure of QA teams, provides an excellent means to support the implementation of an incentive system if such a program is desired. Indeed, there is no reason why the QA&SR program cannot encompass the added dimension of establishing productivity standards using work measurement techniques. There is a natural link between quality and productivity.

Gerald Pacella, executive vice president of the Fidelity Bank's operation department, holds the position that quality and productivity are inseparable. Under Pacella's executive leadership, the QA&SR program supports a consultant's incentive-based productivity improvement program. The QA teams incorporate the incentive standards and profit improvement recommendations made by the consultant team into the program's overall improvement strategies. The philosophy of the QA&SR program is to use whatever proves to be useful in improving the quality, productivity, and profitability of a department.

There are hundreds of large and small consulting firms skilled at developing and installing an almost endless variety of job redesign, work measurement, and wage incentive plans. Even before Frederick Taylor and Frank Gilbreth in the early 1900s, consultants showed management how to increase worker productivity. Millions have been spent on consultants' fees and will continue to be spent. Much of the money and effort will be well invested. Some of the money, however, will not be recovered in improved productivity that will favorably affect the bottom line.

Let's take a quick look at the techniques most frequently used by consultants to improve employee productivity. Try to assess the worth of these techniques while bearing in mind how they can be incorporated into your organization's QA&SR program to increase productivity while improving quality. Consider this: Hasn't it been your experience that the most productive workers are usually also the ones who do the best quality work? And on the other side of the coin, aren't the least productive workers also the ones who make the most mistakes and do the poorest work? If that is your sense of the realities of the work place, then you should be interested in ways to use productivity improvement techniques as part of the quality assurance program.

Job Design

The first and most important step in increasing productivity and improving quality is to look at the way jobs are designed in your organization. It is critical to understand that the character of employees has changed since the 1960s. The nature of most work tasks has also changed, because of the all-pervasive presence of computer systems: mainframe host systems and microcomputers (PCs). These changes make it absolutely necessary to reexamine and redesign jobs to match the temperament and skill level of contemporary employees and the greater complexity and volume of work that must be processed in your organization's departments. Senior management must realize that the sons and daughters of the processing clerks of 20 to 30 years ago are as different from their parents as their parents' typewriters, adding machines, and 803 Proof Machines are from the word processors, desk calculators, CRT terminals, PCs, and electronic reader/sorters of today.

In redesigning jobs to meet the demands of today's employees, office machines, and automated systems, the following factors must be taken into consideration:

- Quality control as part of each employee's job

- Cross-training employees to perform multiskill jobs

- Worker participation in the design of their jobs

- Quality assurance and productivity improvement, focused on employee teams that participate in identifying, implementing, and administering remedial strategies

When redesigning jobs, the objective is to develop work assignments that match the goals and objectives of your organization, the psychological and sociological needs of the employee, and the technological architecture of your organization.

Before moving on to a detailed examination of the job-redesign process, it is necessary to consider the overall working environment. To achieve the full benefits of job redesign, executive management must build a superior management environment that motivates employees to give their best performance. The following policies are essential to building a superior working environment:

1. An informal, honest, open-door communication policy, with no artificial distinctions between management and operative employees. Senior officers should walk through their departments frequently and should be on a first-name basis with their operative employees.

2. A policy of developing and promoting employees from within the organization and a commitment to lifetime training, education, and career counseling to help all employees achieve their full potential within the company.

3. Decentralization, with an emphasis on keeping decision-making responsibility at the level closest to where it impacts. This gives managers, supervisors, and operative employees a sense of control over their jobs and the business they bring in.

4. Management-by-objectives to provide a sound basis for measuring managerial and supervisory performance.

While executive management works to develop and maintain a superior working environment, line managers and supervisors can work to redesign their employees' jobs, using the concepts of job enlargement and enrichment.

Job enlargement consists of adding a wider range of activities to jobs. In essence, it broadens the scope of employees' jobs to make them more interesting. Enlargement is done horizontally to give employees a sense of performing a whole unit of work, a feeling of accomplishment.

Job enrichment is similar in intent to job enlargement, but it is much more powerful and effective in that it adds higher-order activities to jobs—activities such as planning, organizing, inspecting, and controlling. Job enrichment broadens the scope of employees' jobs by giving them accountability for a natural unit of work. A natural unit of work is anything with which employees can identify and which, when completed, gives them a sense of satisfaction.

The aspect of accountability also gives employees some sense of ownership or proprietary interest. Employees are encouraged to form relationships with the people who send in work and those who get completed work from them. This gives employees a sense that their work is important to other people—that they have a responsibility to serve their clients well so that they can do their jobs right, too. Communication networks are established to provide feedback on their client relationships to tell them if their work is satisfactory. Employees are also encouraged to give feedback to the employees from whom they get work. Finally, employees are also given the authority to schedule and plan their daily work. Job enrichment is still a very effective way to redesign jobs to build employee satisfaction and motivation to give good job performance.

Work Measurement

Another way to increase productivity is to study the overall processing system to make it as efficient as possible. Work measurement analyses focus on determining what is done, whether it should be done, whether it is being done in the most efficient way, and how should it be done.

Once it has been determined that an activity is worth doing, several methods are used to find the best way to do it. The analysis may be directed at finding a worker or supervisor whose method of performing the activity is best and then training the other employees to perform the activity in the same efficient way. In some cases, a consultant or analyst may have personal experience in performing the activity in the most efficient way; this consultant or analyst may teach all the employees their method of performing the activity.

A supposedly more scientific way is to study what the activity is intended to accomplish and then use time-and-motion study techniques to determine the best way of performing the task. These techniques date back to the early 1900s and were developed by early industrial engineers such as Frederich Taylor and Frank Gilbreth. They advocated breaking the task into its smallest components and determining the most efficient way to perform each component task. The small tasks were then combined into one larger, efficient procedure.

The Taylor/Gilbreth work method approach is still used by many consultants who are desperately trying to adapt these mass-production, assembly plant techniques to the emerging new market of service companies. The approach consists of decomposing manual tasks into small substeps and attempting to force human beings to perform the series of substeps like programmed robots over and over all day long. A manager does not need a MBA or master's degree in psychology to know this approach is not going to work with today's workers. It never really worked that well even in Taylor's or Gilbreth's day, despite the many efficiency experts still plying their trade today. The myth persists only because the idea of training workers to do their jobs as efficiently as possible has an irresistible allure to management looking for ways to increase employee output.

Having harshly criticized the scientific management approach of Taylor and Gilbreth, let me point out that there are also benefits to work measurement. It makes sense to train employees to perform the activity in the same efficient way as one employee has been observed to use. If a consultant knows how to perform the activity more efficiently than the employees performing the activity, it also makes sense for the consultant or your training director to teach your workers to do the activity the way the consultant advises.

Productivity gains are accomplished by studying the work flow, looking for recurring delays, and eliminating the causes of those delays. Gains in productivity can also be made by improving the layout of the work site and the work stations. Productivity can also be improved by taking telephones off clerical personnel's desks to eliminate distracting telephone calls. These calls can be channeled to one or two client service representatives, who field all calls.

Productivity Standards

An integral part of the work-measurement approach is the setting of productivity standards. This approach consists of separating a job into discrete measureable elements and timing a representative sample of employees—or a very skilled employee—as they perform the job, with all its sub-steps. With repeated observations, you can arrive at the average time it takes to complete the job. A time study used to set a time standard for the completion of a job requires that a representative sample of employees be observed per-forming the job over an extended period of time. The average time needed by the employees to complete each element of the substeps is calculated. Many refinements are used, such as performance ratings to normalize the timing data taken from an individual employee before the timing can be ap-plied to the group of employees doing similar tasks. Load-ings for rest, fatigue, and delays are added to the timings to make the time standard more realistic in what can reason-ably be expected under normal working conditions.

What senior management must understand about work measurement/time study techniques is that the standards are accurate and reliable in providing a measure of what employees are "capable of producing"—and that is all. Getting employees to actually produce accurate, quality work at a rate equal to or close to what they are theoretically capable of doing requires much more than just a reporting system that tracks employee output and compares it to what they are capable of producing. Department heads who have had consultants come into their area, setting up standards and reporting systems, can tell you a lot about their frustration when actual production falls short of standard production. They can try everything they know to get their employees to work at the rate that the consultant says they are capable of working, but nothing works.

Wage Incentive Plans

Let's look at what may be the answer to motivating employees to work at their full capabilities—incentive plans. Incentive plans are based on the belief that for a few dollars more in their paychecks, employees will work as hard and as intelligently as they can for their full shift. It puts a lot of emphasis on money as a motivator for most employees. Let's make the working assumption that indeed money is a strong motivator for most employees. Wage-incentive plans fall into three broad categories:

1.   Plans based on time worked plus work output

2.   Plans based on work output alone (piece rate)

3.   Plans based on general performance of the employee or group of employees over an extended time period

The first plan is the most common, with a basic wage to which an incentive bonus is added based on output above a certain level. The output component is based on productivity standards that are set for the number of work units processed per hour, per day, or per week. A work-measurement approach is used to time a group of employees to see how long it takes them to complete a task; based on that time, a time standard is set for the job. For example, a group of employees will be timed to see how long each of them takes to process a stack of loan applications. Let's say that the average time for the group is 12 minutes per application. To this is added a rest, fatigue, and delay factor of 15 percent, raising the standard to 13.8 or 14 minutes per loan application. An incentive plan is set up to give employees several incentive points for meeting the standard consistently over a period of time, such as one day. More incentive points are given for exceeding the standard. With this approach, employees are encouraged to increase their earnings by working to meet or exceed the standard.

Of course, refinements are added to this bare bones description of incentive systems. In general, however, that is the essence of the idea—employees are paid for what they produce instead of the time they put in. With good management and consistent updating to make sure the standards are fair, the system works.

Piece-rate systems pay employees solely on what they produce. There are only a few companies using pure piecework systems in service industries. Again, if administered flawlessly, such programs will achieve the desired results.

To sum up, a QA&SR program is an effective way to support any management or consultant program intended to improve performance. In many cases, a well-developed QA&SR program can incorporate the objective of the separate program into its process. By doing so, it can often achieve the same objective with more lasting benefits, making the incentive program part of the line management process. Whether or not the QA&SR program has a productivity component that uses an incentive factor is not important. What is important is that as the QA&SR program takes hold and is accepted by departmental management, it becomes integrated into the management process, thereby improving it.

Using Recognition To Gain Commitment

An award program should be established as part of the QA&SR program to recognize and reward managers, supervisors, and employees who have consistently performed outstandingly. People appreciate genuine recognition for good performance, and formal recognition will build support for the QA&SR program.

It is critical that recognition be given for achieving specific goals or other identifiable performance. Employees must believe that awards are deserved. An effective awards program will make employees feel that management needs their help and appreciates it. The awards need not be costly, but they must be given—and received—seriously.

Performance Evaluation

The best way to ensure that QA&SR awards and employees' commitment to the QA&SR program are taken seriously is to make sure that employees' contributions to the QA&SR Program figure prominently in their performance evaluations. The performance evaluation form should contain a section dedicated to describing the employee's contribution or lack of commitment to the QA&SR program. The supervisors and managers who do the evaluating should understand the importance of the QA&SR program and should explain it to the personnel reporting to them.

Supervisors, managers, and divisional officers should also understand that their contributions to the QA&SR program, as well as the contributions of their staffs, are an important dimension of their job performance. This will ensure that everyone knows how serious executive management is about getting everyone's commitment to the QA&SR program.

An effective way to set up a recognition awards program is to design a formal structure to control the selection and evaluation of people for the awards. The best way to ensure effective control of the nomination process is to design an awards nomination form. The form should be completed by the QA&SR project leader. This gives the project leader the power to reward people who contribute to the success of the program, which is important. Additionally, the QA&SR project leader is more able than anyone else to evaluate the contributions made by QA team members and leaders to the program. The power to reward QA team members and leaders is important to the QA&SR project leader. It is the only positive power base under his or her control. The only other power base the project leader has is to be able to report lack of cooperation or support by a QA team member to his or her supervisor, and that power is best used only when everything else fails.

Exhibit 9.3 is a prototype of a nomination form that can be used by your organization.

Exhibit 9.3:
Nomination for QA&SR Award

TO: Chairman, Awards Committee

FROM: QA&SR Project Leader

SUBJECT: Nomination of Candidate for QA&SR Award

The candidate named on this form is hereby nominated for the QA&SR award.

Name Bill Smith

Position Manager

Department & division Trust accounting

Address & extension Vault

Quality assurance team Trust accounting QA team

Reasons why the candidate should be considered for the QA&SR award.
Bill helped to identify several serious problems that were causing errors in accounting statements being sent to trust clients. He worked with the QA team analyst to prepare situational analyses and documentation of the problems. He further researched the problems by meeting with trust managers and officers from other institutions to determine how they managed their operations. Based on his analyses and research, he developed several remedial strategies to correct the problems and the conditions causing them.

A structure should be put in place for identifying candidates for QA&SR awards. This will provide a means for QA team members, the QA team leader, or the QA council leader to identify candidates for QA&SR awards. In order to set up such a structure, it is necessary to set up criteria for selecting candidates. Exhibit 9.4 is a prototype of a bulletin that can be distributed to the QA team members by the QA team leaders for identifying award candidates.

Exhibit 9.4: Guidelines for Electing QA&SR Award Candidates

QA team members should look for candidates in their units who are outstanding quality performers and who, through example, encourage other employees to do quality work.

When looking for likely candidates, consider the following guidelines:

1. People who have made an outstanding contribution to the QA&SR program.

2. People whose work output is consistently recognized as the standard for quality by their peers and/or supervisors.

3. Managers and supervisors whose support and encouragement to their employees create a working environment in which employees achieve the highest quality performance.

4. People who demonstrate, through their personal actions and leadership, support for the QA&SR program and who encourage others to do the same.

5. People who submit numerous suggestions for quality improvements, many of which have been implemented by the company.

Send your nominations to the QA&SR project leader. All nominations will be acknowledged. The nominations will be evaluated by the Quality Awards Committee. Presentations of awards will be made at the awards and recognition dinner.

Cut-off date for all nominations is November 10th.

Monetary versus Nonmonetary Awards

Whether to use monetary or nonmonetary awards depends on management philosophy. If an organization favors productivity-based incentives, it will favor monetary awards. If an organization adheres to the policy that it hires the best people it can, pays them a fair competitive wage, and expects them to perform at their full capacity, it will favor nonmonetary awards.

The advantage of monetary awards is that people who receive the awards are delighted. To some extent, the hope of receiving a monetary award induces them to work on the QA&SR program. However, there is also the school of thought that most people always try to do the best they can and tend to consistently perform at a level close to their best effort. If that is truly the case, then monetary incentives are unnecessary and simply add nonessential cost to the program. Additionally, as the French social commentator La Rochefoucauld pointed out: "For everyone rewarded, a hundred others are disappointed and embittered."

A serious drawback of monetary awards is that people who feel they made significant contributions to the QA&SR program but were not given an award may become resentful of those whom they feel were unfairly favored. Unless you believe that one person's outstanding contribution is more important than the contributions of hundreds of others, monetary awards may not make good business sense.

Nonmonetary awards, on the other hand, enable the company to reward a broader base of people, and it is less likely that anyone will be bothered by a coworker getting recognition. Furthermore, recognition can be widely distributed at low cost—perhaps the price of a lunch or dinner.

The QA&SR awards should be given at an awards dinner hosted by the senior department officer for all the QA team and QA council members. Recognition should also be given in the company magazine. Articles and feature stories in your organization's journal can be used to give recognition to the people and to add to the visibility of the QA&SR program.

Appendix A

Sample QA&SR Program Quarterly Report

A portion of an actual third-quarter report to senior management is reproduced in this appendix to convey a sense of the content and style of the quarterly progress report. It also provides an insight into the quality problems and remedies the QA teams worked on, which can serve as a guide to the kinds of problems and remedies your QA teams may want to address.

The report should begin with an executive summary that presents a brief overview of the accomplishments of the QA&SR program during the quarter. Following the executive summary, there should be an organizational chart showing the QA teams and QA councils matrix organizational structure. If some QA teams have been proposed but are not yet formed, these should be shown on the chart with a notation. The organizational chart helps management develop a sense of the breadth of coverage of the organizational structure.

The organizational chart should be followed by detailed descriptions of the QA teams' actions, accomplishments, and plans. These descriptions should be broken down by department.

In addition to the narrative description of what was done, the quarterly report must contain trend charts that show the effect of the QA teams on key quality measures (error rates and reliability statistics). When appropriate, narrative statements explaining trends can be written at the bottom of the charts. The target line represents the standard set for the key quality measure or service reliability measure. When data on competitors or the industry at large is not available, it is omitted.

At the end of this quarterly report, two figures from an end-of-year report are provided. These are shown as an example of yet another type of graphic representation format that is useful in portraying the impact of a QA&SR program on error and reliability rates. Here also the graphic presentation can be augmented with a narrative where appropriate. Again, when data on competitors or the industry at-large is not available, it is omitted. The target line represents the standard set. It is a reference line against which to compare performance.

QA&SR Program
Third-quarter Report

Executive Summary

The QA&SR program is functioning effectively, with a healthy mix of well-established and newly formed QA teams. The established QA teams meet every three or four weeks, and the newer teams meet weekly or biweekly.

Quality councils have been formed in the securities and check processing operations to give direction, to evaluate, and to integrate the QA&SR process into their management functions. A quality council is being formed in the computer services operation.

Major accomplishments for the third quarter, grouped by operational areas, are discussed below.

Securities

- Error rates were held flat in most areas, except for the coupon and bond-processing departments, in which record volumes strained processing procedures.

- The quality outreach program, in which client financial institutions with chronic quality problems are identified and visited by QA team members, is improving the quality of work sent in for processing.

- New forms and control procedures are also helping to eliminate the underlying causes of errors.

Check Processing Operation

- Error rates showed significant decline in the third quarter, and, overall, this year's error rates are much improved over last year's.

- Several new controls and new forms, such as the Standardized Cash Letter, are improving the quality of work sent in for processing. New control procedures in the preparation and settlement areas are improving the efficiency of downstream, sequentially linked areas such as adjustments.

- Identifying client financial institutions with chronic quality problems and visiting them or discussing the problem over the phone with the appropriate officer is improving the quality of incoming work.

Computer Services

- System up time improved over last year.

- Workshops were held among the QA teams, resulting in improved communication, clarification of functional responsibilities, and improved working relationships.

- Several structures were put in place to ensure improved communications and integration of functions among the four computer services divisions.

ACH, Wire

- Tracking of error counts has begun.

- Enhancements to improve efficiency are being worked on.

- An analyses of ACH tape misroutings was completed, and recommendations are now under study to correct this problem.

Future Plans

- Plans and schedules are being made to expand the QA program into the accounting department in June and into the cash department in September.

Quality Assurance Team Structure
Anthony DiPrimio, Project Leader

SECURITIES OPERATION

Johnson, V.P.

Coupons QA Team
Sackley, Mgr. Team Leader
Kinney, Sup.
Hamilton, Unit Head
Cappe, Unit Head

Consignment QA Team
Och, Mgr. Team Leader
Ward, Sup.
Lombard, Unit Head

Issue QA Team
Sheaffer, Dir. Team Leader
Miskin, Sup.
Garrison, Unit Head
Thomas, Unit Head

Securities Transfer QA Team
Brown, Mgr. Team Leader
Kinney, Sup.
Knight, Unit Head

COMPUTER SERVICES DIVISION

Dolby, S.V.P.

Data Services QA Team
Shel, Officer Team Leader
Cooper, Mgr.
DiMarco, Mgr.
Glas, Mgr.
Gorman, Mgr.

Automation Plan. & Adm. QA Team
Fisher, Officer Team Leader
Smith, Mgr.
Williams, Mgr.
Sheldon, Mgr.

Sys. Develop. QA Team
Flana, V.P. Team Leader
Cross, Mgr.
Lock, Mgr.
Kellman, Mgr.

Technical Support QA Team
Willard, Officer Team Leader
Houselman, Mgr.
Evans, Mgr.
Mason, Mgr.

CHECK OPERATIONS

Duffy, V.P.

Adjustment QA Team
Hogan, Dir. Team Leader
Bongi, Sup.
McClay, Sup.
Maul, Sup.
Murphy, Sup.

Low Speed QA Team
Camp, Sup. Team Leader
Hart, Sup.
Wood
Kille
Bayer

Preparation QA Team
Carter, Sup. Team Leader
Beck, Sup.
Hamilton, Sup.

Shipping QA Team
Bwon, Sup. Team Leader
Daily, Sup.

SECURITIES COUNCIL

Johnson, Chairman
Sackley
Och
Sheaffer
Brown

CHECK COUNCIL

Duffy, Chairman
Hogan
Camp
Carter
Brown

COMPUTER COUNCIL*

Dolby, Chairman
Shel
Willard
Fisher
Flana

*The Computer Council will be formed in the fourth quarter.

This third-quarter report, in broad terms, presents what the fifteen QA teams have accomplished. All QA teams are meeting regularly. The long-term QA teams are meeting every three or four weeks, and the newer QA teams meet more often.

The matrix organizational structure of QA teams now encompasses every division in the bank. Quality councils made up of the senior departmental officer, who acts as the chairman, the division QA team leaders, and the QA project leader are now functioning in every department.

The purposes of the quality councils are to provide direction to the QA teams, set objectives and goals for the QA teams, evaluate the impact of the QA program in the departments, provide a structure for integrating the efforts of the QA teams, and integrate the QA process into the departments' management function.

The major accomplishments for the third quarter are grouped below by department.

Securities Department

The newly formed municipal bonds division QA team is functioning effectively. The QA process was explained to all the operative employees who make up the processing operation. The operative employees are enthusiastic about participating in the process and identified several environmental factors that are causing quality and service problems. The QA team will study these factors and the recommendations made by the operative employees.

The QA team identified measures for the municipal bonds division and is monitoring these measures. The QA team has selected several quality improvement targets and is working to develop corrective strategies.

The new PC-based operating system is functioning effectively. All the operators have been trained and are gaining proficiency in using the new system to settle record-keeping and to print cash letters.

A new transmittal form is being used by a select group of test banks. The form is being expanded to other financial institutions on a progressive schedule.

An analysis of agents with chronic problems is being prepared for a quality outreach effort. The agents will be contacted by phone or invited into the bank to encourage them to improve the quality of documents sent in for processing.

A number-skills program is planned to improve the accuracy of data entry and other clerical functions.

Several trips were made to important customers who have chronic quality problems with the work they send in for processing. These customers were very receptive to the quality outreach effort. Analyses of the work sent in by the customers for processing following the visits showed significant improvement. One customer has been sending in error-free work since the team's visit.

Check Operations

The new standardized cash letter form is now being used by most of the eight banks in the initial test group. A second test group of remote-location banks with chronic cash letter problems has been selected and will soon receive their new standardized cash letter forms. The preparation QA team is monitoring the first group of banks to assess the effectiveness of the new forms in reducing errors relating to cash letters.

An analysis of quality problems associated with the use of dummy cash letters is nearing completion and will form the basis for developing procedures to control the use of dummy cash letters.

Check operations officers and managers met with banks that have chronic problems in the work they sent in for processing and have been able to correct these problems. Direct contact with chronic problem banks will continue and be intensified.

A letter was sent to all questionnaire survey respondents sharing the results of the survey with them.

Several control procedures and system enhancements have been implemented in the preparation and settlement areas that are resulting in improved efficiency in downstream operations, especially in the adjustment area.

Computer Services Department

The major thrust of the QA&SR program in the computer services department is to improve communication among the four divisions and to establish network structures that will ensure the integration of their highly differentiated functional responsibilities. The objective of improved communication and integration of functional responsibilities is being accomplished by:

- Workshop sessions among the four divisions

- Monthly briefing sessions

- Departmental information news briefs (newsletters)

- Liaison QA team members

- Problem-resolution briefings

In the third quarter, workshop meetings were held among the four divisions. The workshops accomplished the following benefits:

- Functional responsibilities were clarified

- Boundary areas were clarified

- Back-up personnel were identified

- Interface relationships were strengthened and made more effective

- Communication networks were strengthened

- Interdepartmental coordination problems were reviewed so all parties gained a better understanding of the problems, their causes, and the constraints that contribute to them

- Empathy was established among the QA team members for the constraints under which all the divisions must function

One of the major coordinating and integrating structures set in place in the third quarter is the monthly briefing session among the divisions. These monthly briefing sessions focus on coordinating the handling of production problems, clarifying new procedures or proposed policies, reporting on new initiatives, and reporting on the status of established initiatives or projects.

An informal survey of the QA teams indicates significant improvement in communication and supportive empathy among the QA team members who participated in the workshops and briefing sessions.

As a means of further integrating the four QA teams and improving communication, liaison team members participate in other team meetings and report back to their teams.

A problem-management system—the Help Desk—is now functioning effectively. All network problems are reported to this central contact point, and daily meetings are conducted to review the problems and resolve them. The problems and their resolution are monitored and tracked by the control system.

Automatic Clearing House (ACH)

Third-quarter activity focused on the problem of having to remake ACH tapes. The major cause for remaking tapes (38 percent of remakes) is that the tapes are not always received by the financial institution. An additional cause is that tapes are sometimes sent to the wrong institution. The team is working with the check shipping QA team and the transportation department to identify and document the reasons that the tapes do not always reach the correct destination. The team is also documenting the cost of these problems, both to the bank and to the customer.

Electronic Funds Transfer (Wire)

Third-quarter activity focused on system enhancements that would eliminate or prevent the data-transmission conditions that contribute to errors or service problems. The QA team is working with computer services personnel to study the feasibility of enhancements proposed by the QA team.

Fourth-quarter Plans

Plans and time schedules are being developed to expand the QA program into the following areas in 1988:

- Accounting, First Quarter 1988

- Cash, Third Quarter 1988

The present 15 QA teams will, of course, continue through 1987, with the more established teams meeting every three to four weeks and the new teams meeting weekly or biweekly. The quality councils will meet on a monthly basis.

Key Quality Measures: Accounting
"As of" Adjustments Per 1000 Entries Processed

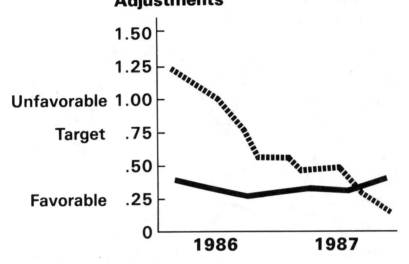

Average Number Account Posting Errors
Per 100 Accounts

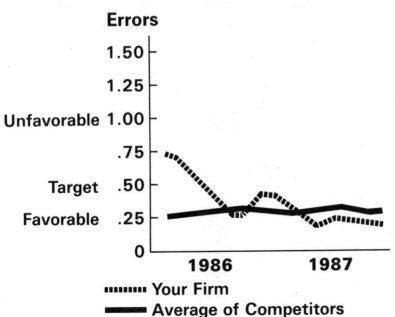

■■■■■■■ Your Firm
━━━━ Average of Competitors

Check Operations

Internal Errors per 100,000 Items Processed

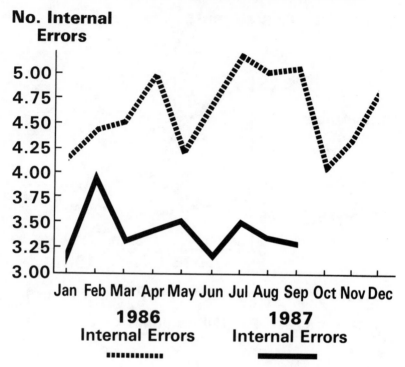

No. Internal Errors

5.00
4.75
4.50
4.25
4.00
3.75
3.50
3.25
3.00

Jan Feb Mar Apr May Jun Jul Aug Sep Oct Nov Dec

1986
Internal Errors

1987
Internal Errors

Check Operations
Internal Errors per 100,000 Items
(1986 versus 1987)

Month	Internal Errors 1986	1987	Volume 1986	1987
Jan	4.15	3.13	1,817	2,108
Feb	4.40	3.93	1,800	2,114
Mar	4.55	3.31	1,768	2,029
Apr	4.98	3.45	2,039	2,247
May	4.32	3.60	1,979	2,104
Jun	4.75	3.17	1,835	2,086
Jul	5.21	3.55	1,938	2,105
Aug	5.09	3.35	1,816	1,999
Sep	5.05	3.31	1,993	2,091
Oct	4.02		2,071	
Nov	4.29		2,113	
Dec	4.78		2,036	
Ytd Tot	56	31	16,985	18,883

Internal errors showed improvement in the third quarter, and this year, overall, is greatly improved over last year. The QA&SR program was implemented in January 1986.

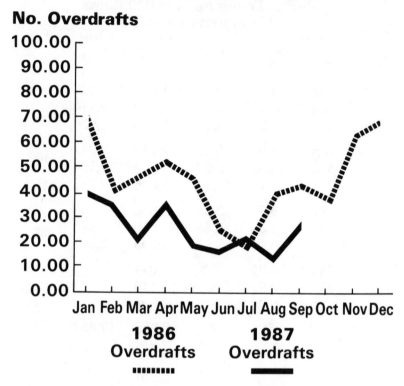

Overdrafts (1986 versus 1987)

Check Operations
1986 versus 1987
Overdrafts

	Overdrafts	
Month	*1986*	*1987*
Jan	70	38
Feb	39	34
Mar	48	19
Apr	54	36
May	48	18
Jun	23	15
Jul	19	22
Aug	36	11
Sep	42	24
Oct	39	
Nov	58	
Dec	63	
	379	217

The overdraft situation has improved during the year and, overall, this year has improved significantly over last year.

Two-year Error-rate Comparison for End-of-year Report

Check Operations
Internal Errors per 100,000 Items Processed

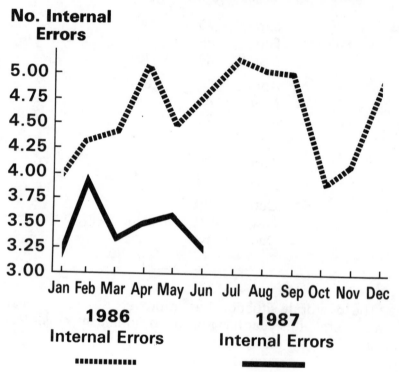

Check Operations
Internal Errors per 100,000 Items
(1986 versus 1987)

Month	Internal Errors		Volume*	
	1986	1987	1986	1987
Jan	4.15	3.13	1,817	2,106
Feb	4.40	3.93	1,800	2,114
Mar	4.55	3.31	1,768	2,029
Apr	4.96	3.45	2,039	2,247
May	4.32	3.60	1,979	2,104
Jun	4.75	3.17	1,835	2,086
Jul	5.21		1,938	
Aug	5.09		1,816	
Sep	5.05		1,993	
Oct	4.02		2,071	
Nov	4.29		2,113	
Dec	4.78		2,036	
Ytd Total	56	21	23,205	12,688

*Checks processed, daily average, in hundreds.

The number of internal errors per 100,000 outgoing items for the first and second quarter of 1986 showed significant improvement over 1987. Volume during this period increased 13 percent.

Appendix B

Developing a Decision Support System

A five-step approach can help your service organization implement a decision support system (DSS) for your QA program. A DSS enables executives to reach decisions on difficult, underspecified quality and productivity problems, when MIS and operations research do not offer enough help. A DSS also integrates and coordinates the efforts of more than one decision-maker working on related parts of a large, complex quality or productivity problem.

A clearer understanding of DSS can be reached by comparing it with management information systems (MIS), which focuses on information and report generation. MIS usually has some inquiry capability and uses a database structured along organizational functions such as production, marketing, personnel or finance. MIS is oriented to, and is dependent on, electronic data processing information flows and data files. MIS is an extension and evolutionary improvement on EDP. In its broadest sense, MIS supports all management activities.

DSS is a specialized management information system that uses operations research and other management science techniques in the form of interactive modeling to support officers' decision-making activities. In addition, DSS communicates and coordinates decision support information among decision-makers at different organizational levels.

DSS is linked to and interacts with other internal information systems. What DSS has in common with all information systems is the goal of improving the decision-making performance of all levels of management.

Intelligence Gathering

A decision-maker goes through a three-phase sequence when using DSS. First is searching for and gathering intelligence related to the problem. In this phase, data are processed and examined for patterns or clues.

The second phase consists of identifying, developing and analyzing alternative courses of action. In this step, the problem is analyzed to understand it more clearly, and solutions are generated and evaluated.

The third phase is selecting a course of action or a solution from those generated in the second phase, then implementing and monitoring it.

DSS is unique in providing the individual manager with the capability to suit his cognitive decision-making style. This gives the decision-maker control over the process.

Consider an example of what DSS did for a service organization. The operations department was frequently asked for quality and productivity analyses to support department officers in controlling problems. The reports usually included comparisons of quality and productivity rates and analyses of backlog and pending work; the department assigned analysts to prepare the reports when they were requested. This company now uses its DSS to track quality, productivity and through-put rates. The monthly reports are reviewed by senior management to spot quality or productivity problems in the inception stage and to take timely action to correct them. Frequent users access the DSS directly. Information that once required weeks for an analyst to work up manually is now available in days. Additionally, the information is more accurate, up-to-date and comprehensive.

Evaluating Need for a DSS

Interest in DSS usually comes from people who recognize a need for more information or quantitative methods to help them reach decisions on difficult, non-recurring problems. They may be line or staff people who have heard about DSS or are using some form of MIS that is inadequate for their needs.

Here are some indications that DSS may be needed in your organization:

- Frequent requests for revisions to computerized systems used to monitor quality and through-put rates. System updates or enhancements may mean managers and officers need more information than they are getting.

- Often-asked-for studies or analyses on recurring topics: these may be quality and productivity related; or operational, marketing or management science oriented. This is not an absolute indication of the need for DSS, but it raises the question.

- Officers and analysts responsible for strategic planning or business planning find that much of the operational information they need is either not available or is in a form that requires a lot of manual analyses to use.

- All levels of management question whether the quality and productivity information being collected and distributed contains what is really needed. Even worse, managers and officers are not using existing reports and make decisions with inadequate information.

Any of these situations is an indication that the decision and policy-making needs of your organization are not being fully met. To see if DSS might help, do a preliminary study to learn the kind of information managers need. Typically, user needs assessments turn up comments like: "We need better information system support for long-range planning." Or, "We're doing a good job of tracking operations, but if we want to analyze problems or take advantage of opportunities, we can only get the information we need by manual analyses."

Most service organization officers feel their information

systems do not give them the information they would like to have to plan effectively. The most-often-heard complaint is that there is hardly any up-to-date information on what the competition is doing or may be planning.

What officers want is ready access to quality and productivity information that lends itself to easy analyses and inferencing. They want systems that are designed to supply information customized to their individual decision- and policy-making needs. Plus, they want systems that go along with the way they use information in reaching decisions or formulating policy. And that's what DSS is all about.

As a pilot effort, start by developing specific decision support systems that provide focused information on quality and productivity data to the chief executive officer and senior executives. This will convey internal and external quality and productivity data, providing a variety of analyses, graphic formats, modeling capabilities and report formats, giving executives and analysts ready access to information that would normally require extensive manual effort and time to obtain. Exhibits B-1 and B-2 are examples of the type of reports that can be generated by a DSS.

A DSS does not provide anything that competent analysts cannot provide—given enough time and availability of data. DSS just makes the same information available *directly* to service organization officers and analysts.

DSS Technical Components

DSS consists of database management software (DBMS), model base management software (MBMS) and dialog generation and management software (DGMS).

The database management software functions to create a database where data is generated, restructured, updated and held available for inquiry or retrieval. The database is dependent on external transaction data and nontransactional data which may have to be obtained by logs, surveys, sampling and other information gathering sources. Some of the data needed are not generally gathered by service organizations as part of their normal operating information cycle.

The DBMS must be able to add, delete and combine data quickly and over a wide range of data management functions.

The model base management software uses operations research and statistical decision models that are integrated with the database. The most common types of models are strategic, tactical, operational and financial planning. The models are created in blocks resembling subroutines. The MBMS allows new models to be created quickly. Also, a decision-maker can access and integrate model subroutines and use them to manipulate data in the database through links between the MBMS and DBMS.

The MBMS functions are similar to the DBMS functions of storing, cataloging, linking and accessing. What DBMS does to data, MBMS does to models.

The dialog generation and management software is the most critical component of the DSS because use of the full capabilities of the DSS depends on it. The DGMS consists of the action language used to communicate with the system through the keyboard and function keys, and the display or presentation language that controls the display screen, graphics, plotters and printers.

A user has some choices on dialog style: he can use a question/answer format, menus, command languages and even a fill-in-the-blanks format. The capability to accommodate varied styles and to present data in a wide variety of formats is an important characteristic of DSS.

Development Strategies

There are three DSS development strategies:

1. Fast with a small investment. This approach has the advantage of being low risk with a high payoff on a short-run basis. It consists of using existing hardware and in-house or purchased software. It is aimed at a highly defined specific information need. The disadvantage is that the newly developed system may have little carryover for other DSS needs.

2. Development in stages. This strategy allows for developing a broader DSS capability. It gives an early payoff with the development of the first specific decision support system and provides a framework for successive systems. Through the iterative process of developing successive specific DSS, new technology is assimilated into the systems building process.

3. Development of a DSS generator. A generator (in programming) is a program or routine that constructs other programs or routines from specifically designed sets of instructions, using specific input parameters and skeletal coding. An example of a DSS generator is one that includes sets of instructions for report preparation; inquiry capability; modeling language; statistical and financial analyses subroutines; and graphic display commands. This takes the longest development time and the biggest investment. It also results in the best integration of all components of the DSS and the best architecture. Some software systems using the DSS generator approach have taken six years or longer to develop. With the present improvements in software, this task should take a shorter period of time now. But it is definitely a multiyear undertaking.

Implementing a DSS

How can you implement a DSS in your service organization? There are five phases.

Phase 1: The DSS Development Team

The first step is to organize a DSS team to manage system development. The size of the team depends on the size of the systems, the time allowed to develop them, the number of people who must input, and the number of people who will be steady users. The responsibilities of the DSS team are:

- To understand the DSS concept and its applications

- To plan the development of the DSS

- To manage the DSS generator, if that approach is used; or to integrate the stages of DSS development, if the staged approach is used

- To help users understand, accept and apply DSS.

The DSS team can be staffed with systems analysts from the computer application departments, programmer-analysts, quality assurance and productivity specialists, operations research analysts, financial analysts, marketing analysts and/or strategic planning analysts. The DSS team can be organized to report to the user's department management or to the management of the computer systems operation. The organizational placement of the team is not critical. What is critical is that the DSS team be sensitive, responsive and committed to meeting the needs of the DSS users.

Phase 2: User Needs Assessment

Here the DSS team interviews all users to get a clear idea of their decision-making and planning needs. These sessions also help the users clarify for themselves what decision support information is needed. The interviews and any needs-assessment studies should be analyzed and set forth in a report and reviewed with the users.

Phase 3: Mission Plan and Communications Network

The DSS team should start the project with a clear understanding of its mission and an effective communication network with users and management. At the start of the project, the question of whether to buy software, turnkey systems or use in-house personnel to develop the software must be studied and a decision made.

Phase 4: Developing Specific DSS

The DSS team then identifies the first specific DSS which will be designed to meet a particular user need. This is followed by a system analysis and design. The identification of the second specific DSS must follow quickly in order to assure that it will be integrated with the first specific DSS. The system design must allow for the integration of all specific DSS. Data flows and interrelationships among organizational levels must be clearly defined.

It is seldom necessary to wait until the first specific DSS is completed before starting on the second specific DSS. Developing the later systems in integrated stages assures that the DSS will be effectively linked together for cross-over functions. For example the first specific DSS may be devoted to quality and productivity information. The second specific DSS may be devoted to staffing and incentive information.

Phase 5: Alternative Approach

If you opt to develop a DSS generator, then the DSS team must identify the necessary generator characteristics. This does not mean trying to identify all the desirable features that the DSS generator should have and attempting to match the list with existing software. That approach seldom works. This is because there is very little chance of finding existing software that matches the list of desired features. Another problem with that approach is that it is very difficult to assess the importance of individual features until users have had some actual working experience with the systems.

A better way is to develop the DSS generator characteristics through an evolutionary process. This is based on an analysis of user needs both before the actual implementation of the first specific DSS and through the testing stages. This approach follows the use of some purchased software and some in-house developed software.

The development of a DSS generator can best be approached using a three-step process. The first step is to identify the generator's objectives. The goals should be stated in terms of the decision-making needs of the users. The DSS generator must adapt quickly to changes in the environment and in decision-making needs.

The second step is to inference what capabilities the DSS

generator must have to meet the objectives identified in the first step. Then, list these as the capabilities that must be designed into the DSS generator.

The third step is to convert these capabilities into specific DSS generator features. These will be expressed in terms of the hardware and software features needed to provide the required capabilities.

A DSS generator should be easy to use and offer access to a comprehensive set of databases providing both transactional and analytical information. The information should lend itself to the users' problem-solving and decision-making needs.

On page 205 are a tabular array and a trend-like chart generated by the DSS developed by the author and Ido Millet for the Operations department of Fidelity Bank (Philadelphia). This measure shows the percent of phone calls abandoned by callers to incoming calls. The "% Need Improve" column shows the percentage of improvement needed to reach the target of 3.00, represented by the baseline in the lower portion of the exhibit. The DSS generates monthly reports for key quality measures which senior management uses to monitor the quality of the department's services.

Pages 206–207 show a DSS-produced list of the issues being worked on by the QA teams. The "Team Code" is the QA team's number; "Minute Date" is the date of the team minutes where the problem was raised; other columns refer to status, type of problem, and provide a brief description.

Quality Assurance
Quality and Productivity Measures 1986

Division: Commercial—funds transfer & communication
QPM: % calls abandoned to calls into wire transfer
QPM code: CFTO8   Target level: 3.00   Alarm level: 7.00

1986 Month	S P	Measure: Calls Lost	Base: Total Call in	QPM: Abandoned	% Need Improve	% from Alarm
Jan.		703	8093	8.69	65	− 18
Feb.		689	7406	9.30	68	− 24
Mar.		573	6523	8.78	66	− 19
Apr.		485	7032	6.90	57	1
May		305	5255	5.80	48	21
June		337	5501	6.13	51	14
July		415	5152	8.06	63	− 12
Aug.		350	4462	7.84	62	− 10
Sep.		364	4819	7.55	60	− 6
Oct.		399	6919	5.77	48	21
Nov.		—	—	—	—	—
Dec.		—	—	—	—	—
TOTAL :		4620	61162	7.48	60 %	− 5 %

Percent Abandoned Calls

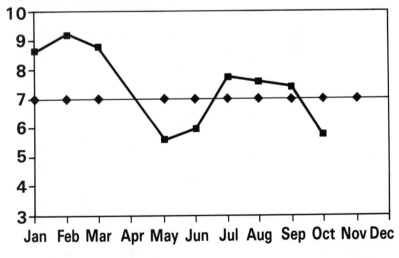

■ **ABANDONED**   ◆ **ALARM**

Issues by Team, Due Date and Priority

Subject: Know/Comm/Qual/Procedures/Task/D.p./Service
Stat: 0:Detect 1:S.A. 3:Pend 4:Exec 7:Abort 8:Merged 9:Comp.

Team Code	Minute Date	Stat	% Comp	Due Date	Subject	Inv.	Bnft	Prio	Sbj1	Sbj2	Sbj3	Situat. Analysis Code
CIN001	03/11/86	3	0	/ /	Need for Motivational and Cross-training Programs	0	0	K		Q		
CIN002	03/11/86	0	0	/ /	Centralize customer inquiries & complaints servicing function	0	0	T		S		
CIN003	03/11/86	0	0	/ /	Communications re: fees to be improved to eliminate improper chrgs	0	0	C		Q		
CIN004	03/11/86	7	0	/ /	Need for better control over Mailroom's routing of mail to Int'l Op	0	0	Q		T	C	

ID	Date				Description						
CIN005	03/11/86	8	100	/ /	Chester misdirects checks to Collections Unit	0	0	0	K	C	Q
CIN006	03/11/86	0	0	/ /	Better Communications needed between Int'l Ops. and Chester Ops.	0	0	0	C	T	
CIN007	03/11/86	8	100	/ /	Investigate Misuse of Interoffice 254 transaction form	0	0	0	Q	P	
CIN008	03/11/86	0	0	/ /	Improperly prepared documentation from Acct Officers and Branches	0	0	0	Q	C	
CIN009	03/11/86	0	0	/ /	Daily work to be randomly reviewed for accuracy (quality control)	0	0	0	P		
CIN010	03/11/86	0	0	/ /	Open lines of Communications between CCL, Credit and Int'l Ops.	0	0	0	C		
CIN011	03/11/86	0	0	/ /	Form Task Force to clarify Commercial Lending Authority Policies	0	0	0	P		
CIN012	03/11/86	0	0	/ /	Training needed for Fidelity Link,LEO,GDK and Domestic Collections	0	0	0	K	D	
CIN013	03/11/86	1	100	/ /	Improve handling of Banker's Acceptanced and Nassau items	0	0	0	Q	P	CIN01

Index